# RECIPES & STORIES
## of Early-Day Settlers

From the Back-Issue Files of
**DISCOVERY PUBLICATIONS, INC.**
with Contributions from Readers

# RECIPES & STORIES
## *of Early-Day Settlers*

From the Back-Issue Files of Discover Mid-America
with contributions from readers.
Compiled by Kenneth C. Weyand


Published by Discovery Publications, Inc.
715 Armour, 5th Floor, North Kansas City, MO 64116

The Publisher wishes to thank the late Ed Elmore and the U.S. Corps of
Engineers, Warsaw, Mo., for providing artwork for the cover. Thanks
also to Sally Schwenk and our many friends at Missouri Town 1855.
A heartfelt thanks to the many writers and contributors whose articles
and research provided not only a guide to early-day cooking, but a rare
glimpse into our heritage. Thanks also to Randall Mertens and
Sam Cangelosi for their production efforts.

This book is available at special quality discounts
for bulk purchases. For details write: Marketing Director
Discovery Publications, Inc., 715 Armour, 5th Floor,
North Kansas City, MO 64116, or call: 816-474-1516.

*Our company also publishes DISCOVER MID-AMERICA,
a monthly newspaper serving collectors and antiquers in
the Midwest, and "On the Collectible Trail" a book
chronicling the origins of American collectibles.*

Copyright © 1988 by Discovery Publications, Inc.
All rights reserved. Printed in the United States of America.


ISBN 1-878496-01-8

## The Woodsmoke Series, Volume I

# STORIES & RECIPES
## *of Early-Day Settlers*

From the Back-Issue Files of Discovery Publications, with Recipe Contributions from Readers

**Compiled by Kenneth C. Weyand**

## *The Stories:*

### HUNTER-TRAPPER DAYS (1800-1850)
Hazards of the fur trade..................................................6
Overnight at Robidoux's..................................................7
Clay County's "Ring Tailed Painter"..................................9
Pioneer Life in Mercer County, Mo.................................11
Wild honey and beeswax...............................................12

### THE SETTLERS GO TO WAR (1850-1865)
The "Kickapoo Rangers"................................................34
Jesse James and the Guerrillas.....................................36
A regiment invades Weston--led by a bear...................37

### REBUILDING THE COUNTRY (1866-1899)
Saloon life was colorful in the 1800's............................48
Passing time in the 1800's meant telling "tall tales".....49
"4-horsepower ferry" served Carrollton, Mo.................51
First fire department in Weston, Mo. proved Murphy's Law............52

### THE 20th CENTURY BEGINS (1900-1929)
Halcyon days of the great showboats............................80
Eating well on the farm--in 1910..................................82
Keeping warm took real effort back in 1914..................86
Memories of an old storm cellar....................................88
I wanted to look like a town girl...................................90

### DEPRESSION DAYS AND WARTIME (1930-1949)
Gathering Greens.........................................................132
Walking to School........................................................137
Another Thanksgiving Remembered.............................139
Judy's Special Christmas..............................................140

# Introducing Woodsmoke

In the early 1970's, I started a scraggly little magazine called *Discover North*, a journal of historical articles and local news pieces that grew into a fairly respectable regional newspaper called *Discover Mid-America*. Some of my more ambitious colleagues suggest renaming it *Discover America*, but that's another story.

Early on, the little paper survived with a lot of help from my friends, who contributed stories and information. I soon began publishing recipes from readers and the two, like peanut butter and chocolate, began to come together in my mind. Why not a book of historical lore, collected from past issues, along with recipes of similar time periods--all compiled into a usable cookbook? In fact, why not a series of books, all with reader participation? The result was *Woodsmoke*, the first in what I envision to be a series of "stories and recipes of early-day settlers."

We have no test kitchens, so although my wife and I have tried some of these recipes, you'll have to take the word of our contributors that they'll work. A few of our contributors are even a little skeptical of some of them--such as the lady who sent in her grandfather's recipe for cooking skunk. My wife and I didn't try this one. And I won't recommend some of the ingredients the old-timers used.

But I think you'll enjoy taking a trip into the lives of the early-day pioneers and later forebears as they "made do" with what they had. And you might add some interest to your own meals by replacing some of the prepackaged entrees with made-from-scratch foods using fresh vegetables, natural spices, and culinary ideas from a century or more ago.

The book is somewhat arbitrarily divided into time periods. Historical purists may argue that the divisions should be different, and I won't argue back. There are many cross-overs, and many recipes have unknown origins. This is essentially a people's book, full of oral histories that have been written down by families. The historical divisions are there to help us remember that progress has changed cooking as well as the other arts and sciences. And it's a long journey from the hearth fire to the microwave.

I hope this book stirs your interest, and inspires you to dig up recipes and anecdotes from your own family history. If you'd like to share them with our readers, include at least three recipes--preferably old-time ones--some information about their background, and the words "OK to use in book or newspaper." Mail them to Recipe Editor, Discovery Publications, Inc. 715 Armour, North Kansas City, MO 64116. If I use any, I'll send you a complimentary book and a 12-month subscription to *Discover Mid-America*.

Kenneth C. Weyand

# Hunter-Trapper Days
## 1800-1850

# Hunter and Trapper Days 1780-1849

## Hazards of the fur trade

As the 1700's gave way to the new century, explorers and fur-traders were beginning to venture west. Their routes were the great rivers, which would be the settling points for the first homesteaders. Trading posts and "factories" were built by William Clark in the early 1800's to stake out the territory and develop commerce with the Indians.

Trapping was the major industry of the period, and many tales are told of these times. One of our contributors, the late Dr. R.J. Felling, wrote of Blackbird, an Oto Indian chief, who would invite an occasional trapper with a richly-laden canoe to be his overnight guest. The meal was dog stew, a popular delicacy of this era. (No, we don't have the recipe.)

Blackbird added an important ingredient--deadly strychnine--to his guest's stew, and the next morning found the trapper buried and the furs in Blackbird's collection. This continued until a trapper, aware of the chief's dubious hospitality, switched

"Fur Traders Descending the Missouri," detail from painting by George Caleb Bingham, about 1845.

bowls with Blackbird. The next day, according to Felling, the village prepared a large funeral for their chief. "A deep trench was dug on the highest hill along the river. Blackbird's body was tied astride his favorite horse and the horse was led into the trench. Everything was covered with a big mound. To this day it is known as Blackbird Hill."

In those days a trapper's worst enemy as he paddled his fur-laden canoe down the river, was not Chief Blackbird, but mice. Attracted by the green pelts, colonies of them would attack, and a year's work could be lost. The solution was to bring along a cat as a traveling companion, and many accounts of early-day trappers mention the cats.

Boys in the river towns would build a thriving business rounding up strays, which they sold to riverboat captains for a quarter. The captains would deliver them to upriver trading posts, sometimes getting more than five dollars for good mousers.

Families who lived in the river towns visited by steamboats often lost their cats to catnappers, Felling reported. "Many pets disappeared," he said, "to later enjoy a more exciting life in the fur trade."

## Overnight at Robidoux's

William Paxton, the well-known chronicler of Platte County throughout much of the 1800's, described "Court at Roubidoux," in which he relates a visit with the founder of present-day St. Joseph. Then a young attorney, Paxton was representing Roubidoux (spelled "Robidoux" in later accounts) in legal matters. The entry is dated November 25, 1839:

"The third term of circuit court was held by Judge King at Faylor's hotel. His next term was for Buchanan, and this I attended. I went up to Roubidoux the evening before court. His house was perched on the hillside. It was of logs on a stone basement. I was shown to my bed on a plank frame in the basement, and was given two blankets. I spread one blanket on the boards, and covered with the other. It was a cold, blustery night, and I nearly froze. In the morning, before day, I heard Roubidoux stirring in the room overhead, and I went up the rude ladder. He asked me in his broken English, French, and Indian how I had spent the night. I told him I had suffered from the cold. "What!" said he, "cold with *two* blankets?" I explained how I had used the blankets. He replied with contempt: "You haven't got even Indian sense, or you would have wrapped up in them."

"The old man had built a roaring fire, and two prairie-chickens and a half-dozen ears of old corn were boiling in the pot. I made a hearty breakfast on these viands. Before court met, I took a survey of the future site of St. Joseph. I saw but two houses: that where I had spent the night and the store above the mouth of the creek. The Blacksnake Hills were romantic. They seemed to be composed of red crumbling earth, with here and there a tuft of grass. From the sides of the hills, at intervals, broke out oozing springs of pure water, which gathered into a bold stream that coursed the prairie bottom to the river. In the rear of the house, on the hill-side, stood four or five scaffolds, supported by poles. On these scaffolds lay the bones of Roubidoux's children. His wives were Indians, and he buried his dead in Indian fashion.

"Court was held in one room and the elevated porch. The docket was short. The most interesting cases were several indictments against Roubidoux for gambling. All the bar except W.T. Wood, the circuit attorney, entered our names in the margin of the docket as for Roubidoux. We got the old man clear on some

quibble, and he was happy. We charged him nothing, but he made all of us pay our tavern bills."

That Robidoux was a gambling man is confirmed in early accounts. Trish Bransky, in her article, "Joseph Robidoux: Trader, Town Builder," which appeared in a now defunct St. Joseph city publication, states that "one (tale) suggests that the fate of our city was dealt at the poker table. It seems that one day in 1839, (the year of Paxton's visit) three businessmen from Independence arrived at Blakesnake Hills and proposed to buy Robidoux's land for $1600. It was their intent to establish a town on the site. Terms were agreed to, and the exchange was to be made the next day. In the evening a card game developed-- and then developed into a quarrel. Robidoux accused the three of cheating him. Whether the charge was valid or merely an excuse Robidoux used after changing his mind about the sale, no one knows; but, because of it, the deal was off and Robidoux later set up his own town."

## Clay County's 'Ring-Tailed Painter,' and other stories

Our very first issue, back in 1973, contained an article by Vera Haworth Eldridge, who related some of the experiences of the first settlers in Western Missouri. In her article, "Early Clay County Settlers," she described

the county in 1820, as newcomers were beginning to arrive from Kentucky and Tennessee.

There were no stores then, just a trading post at Randolph Bluffs where furs and beeswax could be traded for trinkets stocked for the Indian trade. A few miles down the Missouri River, at Fort Osage, settlers could do the same thing. But in order to buy a new axe, a settler had to travel as far as Franklin--many miles to the east--to seek out a blacksmith.

There was no money in circulation then. Travelers carried pelts, a common item of barter. Little currency was needed. Land was going for $1.25 an acre, and the local food supply was wild game that could be had for the shooting.

Mrs. Melinda Houston Estes was 92 in 1901, and the local

newspaper, the *Liberty Tribune*, published an interview that revealed much about life in the early 1800's. Melinda had moved with her family from Tennessee to Old Franklin, Mo. in 1817. She was 8, but could remember well the one wagon and four horses that hauled their family necessities--including "a few utensils, clothing and bedding and her mother's black woman"-- to a log cabin her stepfather had built.

It was a one-room cabin, with a large hearth, offering little more than basic shelter. The chimney, built of wood, was lined with mud, and the fireplace was lined with flat rocks, also held in place by mud. The first winter, the whole family slept in one room, sharing it with their slave family.

In 1821, Liberty, Mo. had only a one-room house built by John Owens, which later became the first courthouse. The next year, William L. Smith built the first store. As she grew older, Melinda's father would take her to the Curtis and Eli trading post at the site of present-day Kansas City to buy calico and blankets.

Early Liberty settlers included Martin Parmer, a rough-cut and independent man who called himself "the Ring-Tailed Painter (panther) of Fishing River." He lived alongside a river-bluff road, where small parties of Indians would travel, begging for food and occasionally stealing horses. After one such incident in 1821, Parmer led a group of neighbors to an Indian encampment, demanded the return of the horses, and killed one of the Indians after a short argument. Other Indians fled and hid in a cabin, but were smoked out, and several were shot. The incident prompted the settlers to build four blockhouses for protection, one of which was built on Parmer's farm. Parmer finally left the "crowded" county for Texas, where he took an active part in its development.

Contrasting with Martin Parmer was Col. John Thornton, well educated and wealthy, who became the county's first judge, and performed its first marriage. He eventually owned more than 3,000 acres of land, and in 1826 established a ferry across the Missouri River. He also built a grist mill on the creek that still bears his name. One of the four blockhouses, described earlier, was built on his farm. His wife, Elizabeth, who was 19 when she married the 34-year-old Thornton, wrote to a friend in Kentucky:

"My husband has made me a home with everything I need, but it is so far away. As I have always liked to see running streams, he had a swath cut in the forest from our home on a hill to the Missouri River for me, so I can see it from my windows on the second floor." The house has been moved to Heritage Village in Hodge Park, west of Liberty, and has been restored.

## Pioneer life in Mercer County, Mo.

Rogers' "History of Mercer County" notes that all the early settlements in northern Missouri were along the streams, in spite of the fact that rich farmland existed on the prairies. Earliest settlers chose the river areas primarily because agriculture in the modern sense was impossible, and convenience to the necessities of life was much more important. The pioneer families depended on wild game for much of their food --especially wild honey. And being close to wood and water meant more to the pioneers than fertility of the soil.

In Mercer County's first decade, the nearest market was Brunswick, located many miles to the south, where the Grand River emptied into the Missouri River, then the "great expressway" for the pioneers. Hauling produce to that point was out of the question, but many local items, such as pelts, beeswax and wolf scalps, could be bartered readily, or turned into cash. In fact, wolf scalps earned a bounty paid by the county, and until the 1850's they were widely used as a medium of exchange.

Another reason for the settlements being on rivers and streams was the fact that many of Mercer County's earliest settlers came from Virginia, North Carolina, Tennessee and Kentucky, areas where hills and forests made up the terrain.

Mercer County was a "border" district, and this retarded its growth, according to Rogers. Neighboring Iowa was free, and Missouri was a slave state. Slave-holding emigrants feared the loss of their slaves if they settled so close to Iowa. Settlers from free states preferred Iowa and Kansas because they were slaveless.

Typical houses in the early settlements were made of hand-hewn logs, containing one room. One log was left out of the cabin for light, and tough bark was used in place of nails. The roof was usually made of clapboards and held in place by rocks. The chimney was sometimes made of stone, but more often of clay and sticks. The fireplace, which served for both cooking and heating, took up one side of the room. Bed posts were set in the floor. Poles extended from them to the log walls, making up the bed. Chairs not brought from the East were made of hickory bark.

Such primitive conditions repel us today, and we assume that the cabins were uncomfortable, but Rogers denies this. For all their crudeness, they were often comfortable and "blessed with old-time hospitality that was of a quality now seldom encountered. People traveled long distances in those days, and houses of public accommodation were few. The pioneers, too, were happy to see strangers and hear news from a far-off civilization. The pilgrim therefore received a hearty and cordial welcome and the best accommodations available. Money was never accepted."

An early settler, "Grandma Loutzenhizer" was interviewed about the early days. "The first winter she and her family spent in the county, their diet was largely confined to corn bread, meat and stewed pumpkin, obtained from the few who lived at a distance. The next fall, the old corn having been exhausted, it was necessary for her to go to the field and get new corn and grate it by hand, in order to have meal for bread. One day a man and his wife came in for dinner. She had already grated her meal and, fearing it would not be a sufficient supply, she mixed it with stewed pumpkin to increase its bulk. She had stewed pumpkin, baked squash, pumpkin bread and pumpkin butter for dinner.

Outside many cabins was a pelt of some kind, tacked up and curing in the sun. "Varmints," such as raccoons, contributed to the pioneer economy. Their pelts could be traded for sugar, molasses, coffee, rice, whiskey or other niceties of life. Or they might be bartered for powder, caps and lead, necessities on the frontier. Rogers states that the interiors of most early-day cabins were "decorated" with a long-barreled squirrel rifle. Using home-made bullets, many an old pioneer defended his home and provided for his family's meat.

## Wild honey and beeswax, and memories of Princeton

Numerous bee trees caught the attention of early settlers in Mercer County. Honey was plentiful, and used in place of sugar in frontier homes. Of greater importance was beeswax, one of the principal exports. Whole neighborhoods would hunt for bees, sometimes going a considerable distance to gather beeswax. Since the forest was largely unmolested, "prodigious quantities" of honey would be found in hollow trees in Mercer County, according to Rogers.

"When I came to Princeton, over fifty years ago," said Jackson Pritchard, an old resident, "there were just two buggies in Mercer County. They were owned by Jackson Pritchard and Seabird Rhea. For ten years after that, if we saw a buggy coming down the road, we could tell who was in it. When there was marketing to do, the whole family came to town. It was a big day. The farmer hitched up his wagon. He and the hired man sat in front, mother and big girls in chairs and the children were stored away in blankets behind. The grown folks browsed among the stores and the children clung to the wagon, in deadly terror of the town boys."

"Father did the selling, took the money and bossed the rest of the family. Mother sat around the stores until noon, and then the whole family ate soda biscuits, cold bacon, minced pie and corned beef at the wagon or lunched on crackers and cheese at the grocery. Finally, after the town dogs and the country dogs had fought it out and the hired hand had made a few trips to the brewery the whole outfit was got together again and pulled back to the farm."

Many times, the family and its wagon would be followed by a drove of hogs that feasted on acorns on the way and would be sold for ridiculously low prices. After the family had done their trading, they would return to the farm with sugar, ammunition and other items that would make their life more worthwhile.

## EARLY-DAY COOKIE
### (Sultana Duncan-Bell, 1804-1882)

My great-great-great grandmother, Sultana C. Duncan-Bell and her family moved into the Platte Purchase area when it opened up to homesteading in 1843, settling near Camden Point. Sweets were not as prevalent then. However, Sultana Bell had this favorite cookie recipe, made up by her help, so that her son, John T. Bell could have them around. I was named for this ancestor.

Using the lard and butter ingredients as specified in the recipe makes a mushy looking dough, but it does make into a nice crisp, well-keeping cookie that will be tasty.

| | |
|---|---|
| 2 cups sugar | 1 teaspoon salt |
| 1/2 cup lard | 1 teaspoon soda |
| 1/2 cup butter | 2 teaspoons baking powder |
| 1 cup milk | 1 teaspoon vanilla |
| 3 teaspoons nutmeg | Add only enough flour to make a soft drop batter |

Work sugar into shortenings. Slowly stir in milk. Then stir in combined dry ingredients. This should be a soft dough. Drop by spoonfuls spaced apart on a greased cookie sheet. Bake in moderate oven until slightly browned and done.

John T. Bell
Lincoln, Neb.

## COOKED LETTUCE

Place a heaping teaspoonful of butter in a 3-quart kettle. Leave it melt. Add 1/4 cup of water. Take 1 onion about 2 inches in diameter and dice it as finely as possible into the pan. Take 8 large celery sticks and cut into as thin strips as possible and dice into pan. Fill pan full with leaf lettuce cut up into 1-inch lengths. Pack it in kettle. Leave it come to a boil over fast heat, then leave simmer over medium heat for 20 minutes. Salt and pepper to taste. Serve as a vegetable. In Denmark they often mix the cooked lettuce with mashed or boiled potatoes. This Danish recipe is supposedly 600 years old.

Maryann Welter
Lee's Summit, Mo.

## CORN SOUP

This is not from an old cookbook but is one that is prepared often in this neck of the woods, and it no doubt goes way back there. I understand my grandmother passed it on to my mom, and thus it came down to me.

4 slices bacon
4 ribs celery, cut in 3" lengths
(1) 16 oz. can or jar cream-style corn
4 or 5 sprigs of parsley
3/4 teaspoon salt

1 medium onion, cut in chunks
1 bay leaf (optional)
1 tablespoon flour
2 cups milk
1/3 teaspoon white pepper
(can use black)

In a skillet, fry bacon until crisp, drain. Remove all but one tablespoon of drippings from skillet. Chop celery and onion and saute in drippings. Add bay leaf, blend corn, parsley, milk, flour, and seasonings for two minutes, then add to skillet mixture.

Cook until mixture comes to a boil and thickens, simmer for 15 to 20 minutes. Garnish with crumbled bacon. Serves two to four.

Maxine Bowles
Knoxville, Ark.

## FARM CHEESE

Take 2 gallons of clabber milk. Heat and strain out the curd as in making cottage cheese. To the curd add:

2 eggs yolks
Butter the size of a walnut

1 level teaspoon soda
1 teaspoon salt

Put in double boiler and stir until curd melts into a smooth batter. Pour in dish to mold. When cold, slice. Will look and taste like cream cheese.

Frances Uttinger
Kansas City, Kan.

## BEST BEET GREENS

When they are very young, these greens can be cooked exactly like spinach, but most people find them more delicious when each cluster of leaves has its root, a tiny beet no bigger than a walnut.

Wash thoroughly in several waters, using salt water first to remove any small insects that may possibly cling to the leaves. Cook in boiling salt water until tender; drain; cut off the tops and then skin the beets by plunging them quickly in cold water and rubbing off the skins.

Drain the tops and mix with the beets and season with salt, pepper and a tablespoon or more of vinegar or lemon juice, and pour over them a tablespoon of melted butter or dripping. If necessary, reheat for a moment in a frying pan.

"This is a recipe used by my grandmother and her family, and is probably 150 years old or more."

Mrs. Donna Miller<br>Kansas City, Mo.

## OATMEAL BREAD

5 teaspoons baking powder
1 teaspoon salt
2 cups oatmeal
1/2 cup milk or water

3 1/2 cups flour
1 tablespoon molasses
1 egg

Cook one cup oatmeal in 3 cups of water until a porridge. Add salt, sugar and cool.

Beat egg. Add milk and egg to cooked oatmeal. Add flour and baking powder sifted together. Mix thoroughly, turn out on molding board and knead well. Bake for one hour in moderate oven.

Frances Uttinger<br>Kansas City, Kan.

## HICKORY NUT MACAROONS

1 egg white
1 cup hickory nuts
1 teaspoon vanilla

1 cup powdered sugar
1 tablespoon flour

Beat the egg white until stiff. Add powdered sugar, nutmeats, flour, vanilla. Drop by spoonfuls on a greased tin and bake in rather hot oven.

Frances Uttinger<br>Kansas City, Kan.

## SOUR MILK DOUGHNUTS

1 egg well beaten
1 cup sugar
1 1/2 tablespoon melted lard
1 teaspoon soda
Salt and nutmeg to taste.

1 cup sugar
1 cup sour milk
4 cups four
2 teaspoon cream of tartar

Add sugar, milk, lard to egg. Mix and sift flour, soda, cream of tartar. Mix together. Roll out to half-inch thick. Fry in hot lard.

Frances Uttinger
Kansas City, Kan.

## MARY BALL WASHINGTON'S BLACK CAKE

1 cup brown sugar
1/2 cup butter

2 squares unsweetened chocolate
1/2 cup milk

Scald and cool. Don't scrape pan.

1/2 cup butter
1 cup white sugar
2 eggs

1/2 cup milk
1 teaspoon soda
2 cups flour

### Icing

1 cup sugar
1/2 cup cream

1/2 cup butter
(Boil to soft ball stage)

Cream butter and sugar, add eggs. Mix in flour, soda with milk. Add chocolate mix. Bake at 350º in three eight-inch pans.

In 1784, Mrs. Washington used one half pound of chocolate. She spelled flour "floure," said "mingle all together," and used "ale baume" for leavening. She baked it in the fireplace for one hour.

Wilmoth B. Dunlap
St. Joseph, Mo.

## KORN CAKE

My aunt, who lived in Atchison, Kan., used this recipe as early as 1904, but it had been in the family many years by then.

Two cups Indian meal,
One cup wheat,
One cup sour milk, one cup sweet.
One good egg that you will beat,
Half a cup of molasses too.
Half a cup of sugar add thereto,
With one spoon of butter new.
Salt and soda each a spoon,
Mix it quick and bake it soon.
Then you'll have corn bread complete:
Best of all corn bread you meet.
It will make your boy's eyes shine
If he's like that boy of mine.
If you have a dozen boys
To increase your household joys,
Double then, this rule I would
And you'll have two korn cakes good.
And when you've nothing nice for tea,
This, the very thing will be.
All the men that I have seen
Say it is, of all cakes, queen.
Only Tindall can explain
The link between corn bread and brain.
Get a husband what he likes,
And save a hundred household strikes.

Mrs. Donna Miller<br>Kansas City, Mo.

## GRANDMOTHER'S BOILED DANDELIONS

Pick the dandelions over carefully. Cut off the roots unless you are fond of their bitter taste. Discard the blossoms also. Wash in several waters until absolutely free from all sand and dirt and then place in boiling water and cook until tender, which will take about an hour unless dandelions are very young. Don't use too much water, as the leaves have a good deal of water clinging to them on account of their many washings, and will mash down when they begin to get soft.

Three cupfuls of boiling water is enough for two quarts of leaves, though if the leaves are tough, you may have to add a little more before they are done. Add a teaspoonful of salt to the water about ten minutes before taking the greens from the stove. Drain off what little liquid there is left into a bowl and save it for a very healthful tonic. It is bitter, but it will do you good to take a tablespoonful occasionally. Chop the dandelions, season with a little pepper, more salt if liked and a dash of vinegar and a tablespoon of melted beef drippings, butter or bacon fat. Put into a frying pan and reheat and serve at once.

If you are extravagant you can cook these greens with a small piece of salt pork or bacon, but in that case you omit the butter or drippings.

"This recipe was hand-written, and found among old papers. This is the way my grandmother cooked dandelions 150 years ago (1830's and 40's). My mother used this recipe also, but I buy spinach in the can."

Mrs. Donna Miller<br>
Kansas City, Mo.

## TEXACUS

### (From Mexican War, 1840's)

This was served to our soldiers who were taken prisoner by the Mexicans. My ancestor liked it and asked for the recipe when he was released.

Mix Together:

| | |
|---|---|
| 1 lb. hamburger | Salt and pepper to taste |
| 1/2 medium onion | 1/2 teaspoon thyme (important) |
| 1/2 cup rice (raw), uncooked | 2 cups water |
| 1 medium head cabbage | |

Carefully take outside cabbage leaves off head and put in hot water to wilt, then wrap large spoonfuls of mixture in leaves and secure with toothpicks, and place in large kettle to boil; put a plate on top to hold down cabbage and keep secure. If you want, put them in pressure cooker; cook about 15 to 20 minutes.

Mrs. Donna Miller<br>
Kansas City, Mo.

## SCRAPPLE

7 cups water
3 teaspoons salt

2 1/3 cups cornmeal
2 cups meat, cooked into small pieces

Make mush by stirring the cornmeal into boiling salted water. Add meat and cook two to three hours in double boiler. Put into a mold. Slice and saute in hot fat.

Scrapple may also be made by using cracklings from which fat has been fried out.

Frances Uttinger<br>
Kansas City, Kan.

*The following recipes were submitted by Carlene Hale, a member of "Friends of Missouri Town 1855." The organization is a support group for a major restoration project by the Heritage Programs & Museums division of the Jackson County (Missouri) Parks and Recreation Department which has created a community of historic buildings near Lake Jacomo, near Kansas City. The project also provides demonstrations of lifestyles and skills of the 1855 era. The recipes are from the 1850s and prior years.*

## ENGLISH PLUM PUDDING

Half a pound of beef suet, half a pound of raisins, half a pound of dried currants, one cup of sour milk, two-thirds teaspoon of saleratus, two eggs, half a nutmeg.

Stone and chop the raisins; the suet should be chopped very fine. Mix in sufficient flour. Some cooks prefer part breadcrumbs mixed with the flour to make it as stiff as cake. Boil three hours.

For sauce, stir together one cup of sugar, half a cup of butter, teaspoon of flour. Thin it with a glass of cold water, boil two minutes. After the sauce is taken from the fire, flavor it with wine or brandy to your taste.

Previous to boiling your pudding, soak the pudding-bag thoroughly in hot water, then cool it, turn it inside out, and dredge it thickly with flour. Pour in your pudding, tie it up tightly, leaving room for it to swell,

& put it in boiling water. Keep the water boiling all the time. As it boils down, pour in more from the hot tea-kettle.

*Breakfast, Dinner & Tea* (1859)

## ROSE BRANDY

Fill a glass jar with fresh rose leaves, pour over them as much white brandy as the jar will hold; cover them & set them by to steep till the flavor of the roses is extracted; then drain them out, filling the jar with fresh rose leaves. Cover them, and let them stand again for at least twenty-four hours; drain them out again, & in like manner fill up the jar the third & fourth time. Then strain & bottle it.
It is thought by many to be superior to distilled rose water for flavoring cakes, pudding, etc. To make the liquid more odoriferous, you may add some fragrant pink leaves, sweet william, etc.

*Kentucky Housewife* (1839)

## CARE OF THE TEETH

Honey mixed with pure pulverized charcoal is said to be excellent to cleanse the teeth and make them white. Limewater with a little Peruvian bark is very good to be occasionally used by those who have defective teeth, or an offensive breath.

*The American Frugal Housewife* (1833)
By Mrs. Child

## BUNCH OF SWEET HERBS

In many of the receipts is mentioned a bunch of sweet herbs, which consists, for some stews and soups, of a small bunch of parsley, two sprigs of thyme, and one bayleaf. If no parsley, four sprigs of winter savory, six of thyme & one of bayleaf.

*A Shilling Cookery for the People* (1856)
By Alexis Soyer

## APPLE SOUP

INGREDIENTS: 2 pounds of good boiling apples, 3/4 teaspoonsful of white pepper, 6 cloves, cayenne or ginger to taste, 3 quarts of medium stock.
MODE: Peel & quarter the apples, taking out their cores. Put them into the stock, stew them gently till tender. Rub the whole through a strainer, add the seasoning, give it one boil up, & serve.
TIME: One hour.

*Mrs. Beeton's Book of Household Management* (1859-61)

## SWITCHEL

Switchel was a very refreshing beverage for the men in the field. No matter how much they drank, it never spoiled their appetite for meals.

1 cup brown sugar
2 quarts hot water
1/2 cup molasses

3/4 cup vinegar
1/2 teaspoon ginger

Stir all ingredients in a crock. Pour into gallon jug and cool.

*Kentucky Housewife* (1839)

## GUMBO SOUP

Prepare a good chicken or lamb soup, stir into it the okra, which thickens and forms the mucilage so pleasant in the soup. Or slice a chicken into shreds, add to them slices of salt pork cut into bits; put them over the fire in water, add butter, spices, chopped celery, onion if you like and thicken with the okra, stirring it a long time.

*Breakfast, Dinner & Tea* (1859)

## CIDER

Put the peelings from the pie apples in a brass pot; add water and boil till the flavor is out of the peelings. Serve cold or hot.

*Breakfast, Dinner & Tea* (1859)

## RUM CATSUP

Chop fine a small handful of thyme, parsley, sweet basil, sweet marjoram, the peel of two fresh oranges & one lemon, half an ounce of mace, half an ounce of black pepper, a quarter of an ounce of cayenne pepper, and a quarter of an ounce of cloves. Put them all in a pan with a quart of good vinegar. Cover it and boil it a few minutes till the flavor of the spices, etc. is extracted. Then strain it, throw in a handful of salt, and stir it until it gets cold; after which stir into it half a pint of Madeira wine and half a pint of rum. Put it up in small bottles, filling them quite full, and securing the corks with leather. Designed to flavor sauces and gravies; they are sometimes sent to the table in castors.

*Kentucky Housewife* (1839)

## CHOCOLATE

In preparing chocolate for your family use, cut off about two inches of the cake to one quart of water; stir it first in a little cold water, till it is soft. Then pour on the boiling water. After it has boiled a short time, add a pint of milk, boil up and serve. Sweeten to serve.

*Breakfast, Dinner & Tea* (1859)

## STRAWBERRY BISCUITS

Bake a soda biscuit, cutting it as large as a dining-plate. Open it while hot, and butter each half well; spread strawberries upon the lower half, sprinkling them thickly with sugar; lay the upper half on, cover it with strawberries, finishing it nicely with white sugar, and eat it warm.

*Breakfast, Dinner & Tea* (1859)

## WATERMELON

Take a fine ripe watermelon, cut out the soft part of it, leaving all the white part of the meat that is firm, and cut them into any fanciful figure you please; put them into strong salt and water, and let them remain in it for ten days; then soak them in fresh water three days, changing the water once a day, and put them into a jar of strong vinegar with a little pepper, mace & cloves. They may be colored green with vine leaves, or red with beet juice, and either makes a pretty garnish for meats.

*Kentucky Housewife* (1839)

## BOILED PEAS

Boil them whole, without peeling, until they are tender, adding to them, when half done, sufficient sugar to sweeten to your taste. This dish is eaten warm for tea.

*Breakfast, Dinner & Tea* (1859)

## POTATO SALAD

Six potatoes  
Six onions  
Two ounces of butter  
Pepper, salt & vinegar to taste

Boil the potatoes and the onions till they are soft; the onions require about as long again as the potatoes. Wipe out the pot in which the potatoes were boiled. Mash the onions in it, slice the potatoes, but don't mash them, and add to the onions. Put in the butter, pepper, salt & vinegar; set it over the fire and stir it till it is hot, when it will be ready for the table.

*The National Cookbook* (1850)

## QUEEN OF PUDDING

Pour one quart of milk over two cups of fine dry bread crumbs. Beat four egg yolks. Add one half cup of sugar, then combine them with one-half cup of sugar. Add a spoonful of flavoring (rose water) and pour into a shallow buttered baking-dish and bake until set in a moderate oven.

Spread the top with a thick layer of fresh or preserved strawberries or raspberries. If fresh berries are used, strew well with sugar. If preserves, use no sweetening. Spread over the top a light meringue made by whipping the four egg whites to a froth with two tablespoonfuls of powdered sugar. Return to moderate oven and cook until the meringue is set and slightly browned.

*The Williamsburg Art of Cookery*

## LEMONADE

To make a quart, take two lemons, or more, according to taste; pare off thinly a little of the rind, or rub lumps of sugar upon them. Squeeze out the juice of the fruit and mix it with two ounces of white sugar, including what has been rubbed upon the lemons. Add the water boiling hot and when sufficiently cool, strain the liquor. This may be diluted with water to the strength required. If lemons are not in season, syrup of lemons may be used, or the crystalized citric acid and sugar, adding a few drops of the essence of lemons.

*An Encyclopedia of Domestic Economy* (1845)
By Thomas Webster

## BROILED HAM & EGGS

Cut the ham in thin slices, take off the rind, wash them in cool water, and lay them on the gridiron over quick coals. Turn frequently, and they will soon be broiled. Take them upon a platter (previously warmed), butter and pepper the ham. Have ready on the fire a pan of boiling water from the tea-kettle; break into it as many eggs as you require for your family, and when "the white" is done, dip out each egg carefully with a spoon, so as to keep it whole, and set it on one of the slices of ham. In that way arrange them handsomely in the dish. Sprinkle pepper over each egg and serve.

*Breakfast, Dinner & Tea* (1859)

## GINGER BEER

Mix together two and a half pounds of sugar, three ounces of good ginger and the grated peel of three lemons; put the mixture into a small cask, and pour on it two gallons of boiling water; shake it up and mix it well, and when it is only lukewarm, put it in the juice of the three lemons with a gill of good yeast, shaking it up again till the yeast is mixed through the other ingredients.

Stop it up slightly and next day bottle it, securing the corks with leather. If you can't procure fresh lemons, mix in two ounces of cream of tartar, which will impart to the beer an agreeable acid taste, and make it very lively and brisk.

*Kentucky Housewife* (1839)
By Mrs. Lettice Bryan

## PEAS & LETTUCE

4 cups fresh peas
1 small lettuce head, shredded
6 spring onions, tops & bottoms, chopped
2 teaspoons sugar
1/4 cup light cream

1 teaspoon salt
2 teaspoons sugar
2 sprigs of parsley, chopped
1 teaspoon salt
2 sprigs of mint, chopped
2 tablespoons butter

Cook peas, lettuce, onions, parsley and seasonings in very little water until just tender. Do not overcook. Drain most of the liquid and stir in cream and butter. Taste to correct seasoning. Additional chopped parsley, mint and onion may be used to garnish.

*Kentucky Housewife* (1839)
By Mrs. Lettice Bryan

## RUSKS

INGREDIENTS: To every pound of flour allow 2 ounces of butter, 1/4 pint of milk, 2 ounces of loaf sugar, 3 eggs, 1 tablespoonful of yeast.

MODE: Put the milk and butter into a saucepan and keep shaking it round until the latter is melted. Put the flour into a basin with the sugar. Mix these well together and beat the eggs. Stir them with the yeast to the milk and butter, and with this liquid work the flour into a smooth dough. Cover a cloth over the basin and leave the dough to rise by the side of the fire; then knead it, and divide it into 12 pieces; place them in a brisk oven, and bake for about 20 minutes. Take the rusks out, break them in half, and then set them in the oven to get crisp on the other side. When cold, they should be put into tin cannisters to keep them dry.

TIME: 20 minutes to bake the rusks, 5 minutes to render them crisp after being divided.

Book of Household Management (1859-61)<br>By Mrs. Beeton

## ROSE WATER FOR CAKES & PUDDINGS

"Gather the leaves (petals) of roses while the dew is on them, put them into a wide-mouth bottle, and pour over some alcohol*. Let stand till ready for use."
(*Use grain alcohol or 100 proof vodka.)

Book of Household Management (1859-61)<br>By Mrs. Beeton

## SPOON BREAD

Stir one cup of corn meal into one pint of boiling water which contains one half teaspoon of salt. Stir one minute, remove from fire and add two tablespoons of butter. Beat well, add four beaten eggs and beat in one cup of cold milk. Beat again and pour into hot buttered baking dish. Bake twenty-five minutes in hot oven and serve from baking dish.

*The Williamsburg Art of Cookery*

## CIDER CAKE

Flour six cups; sugar 3 cups; butter 1 cup; 4 eggs; cider 1 cup; saleratus 1 teaspoon; 1 grated nutmeg.
Beat the eggs, sugar and butter together, and stir in the flour and nutmeg; dissolve the saleratus in the cider and stir into the mass and bake immediately in a quick oven.

Dr. Chase Recipes (1867)

## STEW SOUP

INGREDIENTS: 2 pounds of beef, 5 onions, 5 turnips, 3/4 pound of rice, a large bunch of parsley, a few sweet herbs, pepper and salt, 2 quarts of water.

MODE: Cut the beef up in small pieces, add the other ingredients, and boil gently for 2 1/2 hours. Oatmeal or potatoes would be a great improvement.

*Book of Household Management* (1859-61)
By Mrs. Beetor

## DANDELION

Pick and wash your dandelion and cut off the roots. Drain it, and make a dressing of an egg, well beaten, a half a gill of vinegar, a teaspoonful of butter, and salt to the taste. Mix the egg, vinegar, butter and salt together, put the mixture over the fire, and as soon as it is thick take it off, and stand it away to get cold.

Drain your dandelion, pour the dressing over it and send it to the table.

*The National Cookbook* (1850)

## DRIED APPLES FOR PIES

Pick and wash them well. Then pour over boiling water, enough to cover them. Let them stand all night to soak. In the morning put the apples with the water they were soaked in into your stew-pan. If they have absorbed all the water and are nearly dry, add a little more, simmer them slowly, but do not let them boil. When perfectly soft, pass them through a sieve, and prepare them for pies according to the directions given for apples which have not been dried.

*The National Cookbook* (1850)

## COFFEE

As a substitute for coffee, some use dry brown bread crusts, and roast them. Others soak rye grain in rum, and roast it. Others roast peas in the same way as coffee.

None of these are very good; and peas so used are considered unhealthy. It may be worthwhile to use the substitutes or to mix them half and half with coffee. But after all, the best economy is to go without.

The American Frugal Housewife (1832)<br>Mrs. Child

## TO PLANK A FISH

Lightly butter fillet of fish on both sides. Fasten fish with nails to a heavy oak or maple plank about 12x24. The plank is placed upright against the sides of the fireplace, allowing heat from hot coals to cook the fish.

WARNING: Don't allow fish to remain on the board standing too long or it will fall off. As soon as it flakes to touch, remove the plank from standing position. It is hard to say the length of time needed. Approximately 10 to 15 minutes.

The American Frugal Housewife (1832)<br>By Mrs. Child

## STEWED CARROTS

Ingredients: 7 or 8 large carrots, 1 teacupful of broth, pepper and salt to taste; 1/2 teacupful of cream, thickening of butter and flour.

Mode: Scrape the carrots nicely; half-boil, and slice them into a stewpan; add the broth, pepper and salt, and cream; simmer till tender, and be careful the carrots are not broken. A few minutes before serving, mix a little flour with about one ounce of butter; thicken the gravy with this; let it just boil up, and serve.

Time: about 3/4 hour to parboil the carrots, about 20 minutes to cook them after they are sliced.

Mrs. Beeton's Book of Household Management (1859-1861)

## TO MAKE YEAST

An excellent substitute for this article may be gained from composition of potatoes. Boil and peal some of a mealy sort, and mash them fine, adding as much water, or ale, as will reduce them to the consistency of common yeast.

To every pound of potatoes, add two ounces of coarse sugar, and when just warm, stir it up with two spoonfuls of yeast. Keep it warm till the fermentation is over, and in twenty-four hours, it will be fit for use. A quart of yeast may be thus made from one pound of potatoes, which will keep three months. The sponge should be set eight hours before the bread is baked.

*Modern Domestic Cookery* (1817)

## MOUNT CLARE PEACH CORDIAL

1 dozen ripe peaches
2 cups sugar
2 quarts boiling water

1/2 gallon brandy or apple brandy
1 handful peach kernels
1 teaspoon allspice

Scald peaches in boiling water for 1 minute, then cool quickly in ice water. Pull and halve. Crack peach stones and take out the kernels. Put peaches into a large stone jar with sugar and pour the boiling water over them. Add brandy or apple brandy, one handful of peach kernels and allspice. Shake them frequently. Let stand for 4 to 6 months.

From the mansion of Margaret and Charles Carroll
Baltimore, MD (1763)

## CHICKEN PUDDING

Cut one into eight pieces, half a pound of bacon. Cut into slices; season with one teaspoonful of salt, half of pepper, two of chopped parsley, a little thyme, and one captain's biscuit, well broken. Fill the pudding with the meat, add half a pint of milk. Boil for one hour and a half, serve with melted butter over, and chopped parsley on the top.

*A Shilling Cookery for the People* (1856)
by Alexis Soyer

## STRINGED BEANS

These beans require more or less boiling, according to their age; if young, fifteen or twenty minutes will suffice, but when full grown, half an hour is necessary. Season well with butter, salt and pepper.

*Breakfast, Dinner & Tea* (1859)

## SWEET POTATO BALLS

Get fine large potatoes, boil them, peel & mash them fine, & rub them through a collander. Add to the pulp a little butter, sugar, nutmeg & cinnamon, work it well together, make it into small balls. Lay them on buttered tin sheets, & bake them a light brown in a brisk oven. Put them on the platter surrounding the goose.

*Kentucky Housewife* (1839)

## PORK SAUSAGE

Having removed the skin etc. from a nice, tender part of fresh pork, beat it exceedingly fine, with one-fourth its weight of the leaf fat. Prepare some sage leaves, by drying and rubbing them through a sifter, season the meat highly with the sage, salt, cayenne, mace, powdered rosemary, grated nutmeg and lemon. Work it with your hand till it is very well incorporated, making it a little moist with water. Stuff it into skins, which have been neatly prepared, and soaked in vinegar and water for a few hours; hang them up, and smoke them, and when you make use of them, cut them into links, and stew, fry or broil them.

From *Kentucky Housewife* (1839) by Mrs. Lettice Bryan

## SAUERKRAUT

The heads of white winter cabbage, after removing the outer leaves, are to be cut into fine shreds and spread out upon a cloth in the shade. A cask which has had vinegar in it is to be selected, or if none can be had, the inside should be rubbed over with vinegar or sauerkraut liquor. A layer of salt is to be put in the bottom of the cask, caraway seeds are to be mixed with shreds of cabbage, and they are to be packed in the cask to the depth of four or six inches, and layers of this kind, with salt between each layer, are added till the cask is full, stamping them down with a wooden stammer, as they are put into half their original bulk. Some mix a little pepper and salad oil with the salt. Some salt is to be put on the top, and some of the outside leaves of cabbages. About two pounds of salt are necessary for twenty middle-sized cabbages.

The head of the barrel is to be placed upon the cabbage leaves, and must be loaded with heavy stones, and a common method is for a man who has clean wooden shoes on to tread the cabbage down in the cask. A fermentation will take place and some juice will be given out, which is green, muddy and fetid. This rises to the surface and is to be replaced by fresh brine. When the fermentation is over, the casks are closed up.

*An Encyclopedia of Domestic Economy* (1845)
by Thomas Webster

## MISSOURI CORN CAKES
### (without eggs, milk or yeast)

Sift three pints of cornmeal, add one teaspoon of salt, one tablespoon of lard, one teaspoon of dissolved soda. Make it into a soft dough with one pint of cold water, then thin it gradually by adding not quite one and a half pints of warm water. When it is all mixed, beat or stir it well for half an hour, then bake on the griddle and serve hot.

*Breakfast, Dinner & Tea* (1859)

## JUMBLES

One pound of sugar, three quarters of a pound of butter, one pound of flour, five eggs, one tablespoonful of rose-water. Beat the butter and sugar till smooth and light. Whisk the eggs, stir them into the butter and sugar, then add the rose-water and flour. Roll the dough in strips half an inch wide and four inches long, join them at both ends so as to form rings, sift sugar over, place them on tins, and bake them in a slow oven.

*The National Cookbook* (1850)

# The Settlers Go To War
## 1850-1865

# The Settlers Go to War 1850-1865

## The 'Kickapoo Rangers'

The decade of the 1850's was a time of great tension and bitterness in our country, barely a lifetime old. Politics had gotten bitter, with denunciations and rancor on both sides. Abolitionists and freestaters railed at each other with equal fervor. In the process, newspaper shops were vandalized, families split apart, and whole communities and towns were sacked.

The late Dr. R.J. Felling, relating historical events of the 1850's in the February, 1983 *Discover North*, stated that in 1854 a group of Missourians from Weston, calling themselves the "Kickapoo Rangers," found a cannon that once had been used to protect wagon trains on the Santa Fe trail, and used it against the city of Lawrence, Kansas in what was called the Wacharusie War. The people of Lawrence were kept in a state of seige for several days while the Rangers wrecked the Free State Hotel, which housed a newspaper espousing the abolutionist cause. They dismantled the press and threw it into the Kansas River.

A Kansas historian of the era called the Rangers "a party well-mounted and well-armed that looked as desperate a set of ruffians as were ever gathered together. They still carried the black flag and their cannon, "the Kickapoo." Guns, swords and carbines were decorated with the black emblems of their murderous intentions."

The Rangers saw to it that people sent over from Missouri would be allowed to vote in the Kansas election in an effort to make it a slave state. Their actions so aroused the Free State men that they raised a company of 300 to 400 men, who marched into Kickapoo, a town site at the edge of the Kickapoo Indian reservation near Fort Leavenworth, and stole the cannon.

When Kansas became a state in 1861, Col. D. Anthony was the first to hear the news via telegraph, and the cannon was rolled out and fired to celebrate. Territorial statutes, adopted from Missouri laws, had been used in drafting the Kansas constitution, and the Free State men hated them. Before the cannon was fired, Col. Anthony ripped the board covers from the statute books, rolled them into a cylindrical wad, and stuffed them into the cannon barrel. Someone in the crowd yelled "Send them back their damn laws," the fuse was lit, and the "bluffs echoed for miles" with the explosion. Interviewed later, the colonel said, "I always had a curiosity to know where those statutes really landed."

## Neighbors choose sides

On January 25, 1859, W.M. Paxton noted in his *Annals of Platte County*: "Dr. John W. Doy was apprehended in Kansas, for aiding runaway slaves to escape. He was brought by pro-slavery men to Weston, examined before a justice of the peace, and committed to Platte City jail. He was removed to St. Joseph, and placed in charge of Jailer Brown. A party of his friends appeared at night, with (as they pretended) a noted prisoner to put in jail, and by artifice and intimidation got possession of the keys, and bore Doy away in triumph."

On August 20, Paxton reported that "Geo. P. Dorris advertises "cash for negroes to take south." He builds a dungeon on his farm to confine them."

By the next year, the country slipped closer to war. On January 4, Paxton wrote, "A financial panic is brewing, in view of the expected war between the States. Gold is hoarded, bank bills are discounted, silver is going out of circulation."

On March 26, Paxton wrote, "Rebel flags were flying everywhere, and the national banner was discarded. On the Swaney building, where the *Argus*, the *Tenth Legion*, and the

*Conservator* were printed, a rebel flag floated all summer, much to the delight of Secessionists and chagrin of Unionists. On one occasion Chas. B. Wilson hung from a window of the court-house a national flag, and Tom Dorriss and others tore it down. It engendered some feeling against Dorriss, and he left for St. Louis."

R.S. Bevier, in his "History of the First and Second Missouri Confederate Brigades," recalled the beginnings of the war on Missouri soil, and quoted an address given by the Confederate Congress in an early address to its people. "Instead of a regular war, our resistance of the unholy efforts to crush out our national existence is treated as a rebellion, and the settled international rules between belligerents are ignored. Instead of conducting the war as betwixt two military and political organizations, it is a war against the whole population. Houses are pillaged and burned; churches are defaced; towns are ransacked; clothing of women and infants is stripped from their persons; jewelry and mementoes of the dead are stolen; mills and implements of agriculture are destroyed; private salt-works are broken up; the introduction of medicines is forbidden; means of subsistence are wantonly wasted to produce beggary; prisoners are returned with contagious deseases; the last morsel of food has been taken from families who were not allowed to carry on a trade or branch of industry..."

In another section of the address, the Congress stated, "Missouri, a magnificent empire of agriculture and mineral wealth, is today a smoking ruin and the theatre of the most revolting cruelties and barbarities."

## Jesse James and the guerrillas

During the summer of 1863, Jesse James worked on the family farm east of Kearney, Mo. helping his mother. Brother Frank had joined Quantrill's organization, and participated in the raid against Lawrence, Kan.

According to accounts written in the 1880's, federal militiamen rode up to the Kearney farm, seeking Frank. Before the day was over, they had hanged Dr. Samuel, Jesse's stepfather, and brutalized Jesse's pregnant mother, ripping her clothing and throwing her against a wall. They apprehended Jesse in a field, yanking him away from his plow, and drove him ahead of their horses, lashing him with whips and bayonets.

Jesse's mother managed to nurse Dr. Samuel back to health. Two weeks later, the militia returned. But by this time Jesse was gone. He had joined "Bloody Bill" Anderson, one of Quantrill's most feared lieutenants.

By the following year, both Jesse and Frank were riding with Anderson. Among their accomplishments that year was the defeat of a Union force near Centralia, Mo. By September, the guerilla leaders organized a 200-man army and prepared to raid Columbia. On the morning of Sept. 27 they rode into Centralia to read newspaper accounts of General Price's invasion into Missouri from Arkansas. While there, according to local reports, they got drunk on stolen whiskey, looted stores and homes, stopped a train, herded 23 unarmed Union soldiers down onto the station platform and killed them, set the train depot on fire, sent the train hissing down its tracks without an engineer, took an Iowa soldier prisoner to swap for one of their own men, and finally rode out of town.

## A regiment invades Weston
### ...led by a bear

The 12th Wisconsin Regiment of the U.S. Army was dispatched to "guard" Weston, Mo. in the winter of 1862, and a force of nearly 1,000 men took over the town, according to Dr. R.J. Felling, writing in an early *Discover North* edition.

At that time, Weston was a railhead, an extension having been completed from St. Joseph less than a year before. But of

"Bloody Bill" Anderson, a brutal and vicious Confederate guerrilla commander. Both Frank and Jesse James were with Anderson when the band wiped out a large Union patrol near Centralia. (Photo courtesy The State Historical Society of Missouri, Columbia)

Missouri railroads at the time of the Civil War

greater military significance was Weston's proximity to Fort Leavenworth, and the fact that it was in Platte County, whose loyalty to the Union was questionable at best.

The soldiers had traveled by rail from Madison, Wisconsin by way of Chicago, then down to Quincy, Illinois on the C.B.& Q. From there they walked 22 miles in sub-zero weather to the Mississippi River bank opposite Hannibal, where their officers had intended them to cross on the ice. But the frozen river was not strong enough to hold, so a channel was cut and the troops ferried across on a steamer. From Hannibal they were loaded into a freight train which had seats made from planks laid crossways so close that they men sat with their knees touching the planks in front. They endured two days of this, with no fire or warm food, and the temperature registering 20 below. When the train stopped for water, the men threw out the planks and replaced them with straw from a nearby farm. They arrived in Weston about noon January 15, 1862.

With great relief the regiment struggled out of the freight cars and into the below-zero cold at Weston, where they stomped in the snow and waved their arms to warm themselves. A bugle sounded to "form ranks," and a dark figure was pulled from the boxcar and brought to the head of the column, to begin the march up Main Street into Weston. The dark figure turned out to be the most comfortable member of the regiment--it was a black bear.

The bear was housed in a large dry-goods box, and had the liberty of a 50-foot rope. One of the men would wrestle the bear, much to the delight of the local children. Harlan Squires, the bear's official keeper, would take the bear for walks along Weston's streets, followed by a parade of barking dogs and giggling children--both black and white. The bear liked hardtack as well as the men in the regiment, but he had a special craving for apples.

Although the soldiers were admonished not to loot or destroy property, a few chickens and turkeys found their way into stewpots, and the little German bakery on Main Street found itself missing a few pies and cakes on more than one occasion.

One of the younger privates had noticed turkeys in a back yard as he passed down a side street on guard duty, but they always eluded him. He had nearly given up hope of foraging, when he saw a good sized fowl perched on the fence. Drawing his bayonet, he struck a fatal blow. Thinking of roast turkey, he grabbed his prize, stuck it under his coat and started for quarters. He had gone only a short distance when he glanced down and saw the tail of his bird stretching out nearly six feet behind him and brilliantly shining in the moonlight. He went to the creek and carefully shoved the tell-tale evidence through a hole in the ice. Then he went to bed and kept his secret. But he couldn't help wondering how roast peacock would have tasted.

## BLACK GEORGE CAKE

3 tablespoons sugar
1 cup dark molasses
2 1/2 tablespoons butter
1 egg yolk
2 cups all purpose flour

1 teaspoon cinnamon
1 teaspoon cloves
1 teaspoon soda
1/2 teaspoon salt
1 cup boiling water

Mix all together and bake in two layers in 350º oven.

FILLING:

1 cup raisins (ground up)
4 tablespoons water

1 cup sugar
1 egg white

Boil water and sugar until it spins a thread. Add raisins and pour slowly over well beaten egg white. Spread on cool cake.

"When I clipped this recipe out of a newspaper 35 years ago, it was said to be more than 90 years old then. I usually frosted with a chocolate frosting."

Lulua Fillmore
Independence, Mo.

## MISSOURI COUNTRY HAM

Missouri settlers brought with them a taste for many southern dishes which they in turn refined in their own way. One of these is the "Country-cured Ham." Virginia Hams tend to be salty, and those from Kentucky were hickory smoked. Settlers experimented with the sugar cure and came up with a ham that has some of the flavors retained from the south but milder and not as dehydrated and tough. It is very tasty and universally preferred by those who try it. Many central Missouri brands can be found at the local supermarkets packed by the slice and they are very good.

Place ham in cold skillet (heavy is best, such as cast iron). Cook slowly to brown with the fat intact, and cover with a lid. When brown, turn to brown other side. After both sides are browned, remove to a warm plate and remove fat, before serving. Use grease and drippings for gravy. See below.

### "Red Eye" Ham Gravy

Bring ham grease and small amount of water to a boil. Scrape bottom and sides of pan to include drippings in water. When boiling add approximately one cup of black coffee to water and stir until back to a boil. Serve with ham and biscuits. (Sometimes, boiled corn grits are served with this gravy.)

Joyce Day Baker<br>Odessa, Mo.

## OLD FASHIONED DUMPLINGS
### (For large rich stewing hen)

When Sherman made his march through the South during the Civil War, food was scarce and soldiers were often fed with what could be obtained locally with few ingredients, and those that would stretch the farthest. This old family recipe is not found in cookbooks.

Cook hen slowly on top of stove in dutch oven with a lot of seasoned water. When hen is done, remove from pan and bring broth to a boil. If hen is extra rich you may have to add more water so that you have the kettle almost full.

Sift together in a bowl 3 cups of flour and 1 teaspoon of salt.

Slowly pour in some of the boiling hot chicken broth until it is the consistency to roll out on a floured board. Mix quickly with a fork while hot. Keep the broth at a rolling boil. Roll out dough on the board to less than 1/4 inch thickness. Cut into 1 1/2 inch squares and carefully drop into the boiling broth one at a time. Let boil good. Turn down heat, put on lid, and cook on simmer for approximately 15 minutes. Serve in bowl with remaining broth.

Joyce Day Baker<br>Odessa, Mo.

## SOUTHERN PECAN PIE

Pecan pie was usually a treat and reserved for holiday meals. Nutmeats were obtained from the woods and carefully cracked and picked in anticipation of the celebrations.

| | |
|---|---|
| 1 cup granulated sugar | 1/2 cup corn syrup (prefer dark) |
| 1/4 cup butter, melted | 3 eggs, well beaten |
| 1 cup pecan halves | 1 unbaked 9-inch pie shell |

Mix sugar, syrup and butter. Add eggs and pecans. Fill unbaked pie shell with mixture and bake in moderately hot oven (400º) for ten minutes. Reduce heat to a moderate (350º) oven and continue baking for 30 to 35 minutes. Cool before serving.

Joyce Day Baker
Odessa, Mo.

## HOW TO COOK A SKUNK

When I was 12 or 13 years old I began to realize cooking could be fun. I wrote to my grandfather, then 81 and a Civil War vet, in a home for vets in Little Rock, Ark., asking for ideas. Here's one I've "walked around" for years, but could never try. I'm 68 now (1988) and don't expect I'll ever try it. Grandfather's letter to me (in his own words):

"I recall a feller worked for me saying of all the wild meat he ever ate skunk was the sweetest meat. Now I was willin to take his word for it, with out proven it. Me, I couldn't get past the idea to try it. I reckon they aint no reason why skunk meat shouldn't be as good as any."

"Skin clean, remove scent glands under front and hind legs. Put in strong salt water and boil about 20 minutes or so. Drain off this here water, add fresh and seasons: pepper, bay leaves, sage. Steam till tender. Larpen' good eaten! Baked sweet tater & wild greens go good with yer skunk."

If you try it, let me know how it turns out!

F. Maxine Adams
Fulton, Mo.

## LADIES' CABBAGE

Boil a firm white cabbage 15 minutes, changing the water then for more from boiling teakettle. When tender, drain and set aside till cool. Chop fine and add two beaten eggs, a tablespoon of butter, pepper, salt, three tablespoons of rich milk or cream. Stir all well together, and bake in a buttered pudding dish until brown. Serve hot.

This dish (from an old recipe) resembles cauliflower and is very digestible and palatable.

F. Maxine Adams
Fulton, Mo.

## POTATO ROLLS

1 1/2 quarts flour
1/2 cup sugar
1 cup potato water
1/2 cup butter and lard
1 tablespoon salt

1 cake yeast
1 cup mashed potatoes
1 cup milk
Yolk of two eggs

Dissolve yeast cake and one teaspoon sugar in lukewarm potato water. Add potatoes, egg yolks, balance of sugar, milk and enough flour to make a soft batter. Let rise until light, then add butter, lard, salt and flour to make a soft dough. Let rise two hours, form into rolls, and bake.

This recipe was printed and distributed by "Queen of the Pantry Flour," milled since 1869 by the Waggoner-Gates Milling Co. in Independence, Mo. Naturally, the original recipe called for Queen of the Pantry Flour. The president of the company occupied the Bingham-Waggoner estate, now a major visitor attraction. The original recipe books are on display at the estate.

Jeanie White
Independence, Mo.

## YEAST BISCUITS

1 cake compressed yeast
Enough flour to make a thin batter

1/2 gallon buttermilk

Let stand in refrigerator to use as needed. Will keep 2 or 3 weeks. To make the biscuits, take any standard biscuit recipe calling for soda, baking powder, salt and shortening, and use this batter instead of other liquid. Soft dough makes better biscuits than stiff dough. Grease pans generously, cut out biscuits. Turn over in grease, so that tops are greased. Bake in hot oven and serve immediately. Not necessary to stand for raising; the heat does that. You'll never be satisfied with other biscuits after you've eaten these.

Doughnuts and dumplings may also be made from this dough. These should be allowed to raise until double in bulk. Drop dumplings in meat stew or bake with fruit. Variation: Use 1 cup of All Bran or oatmeal as a substitute for 1 cup of the flour. This is a very old recipe from the South that my mother saved.

Betty Laverty
Gladstone, Mo.

## GRANDMA PUGH'S CHOW CHOW

1 peck green tomatoes
8 large onions
1 cup of salt

1 large head of cabbage
3 peppers (mangos)

Grind the above ingredients. Mix and let stand overnight. Drain well. Add:

2 quarts vinegar
1/4 pound mustard seed

2 cups brown sugar

Next, tie the following in a cloth bag:

2 tablespoons cinnamon
2 teaspoons ginger

2 tablespoons cloves
1/4 teaspoon cayenne pepper

Boil all together for 30 minutes. Remove bag, seal in hot jars. This recipe came from Kentucky in the mid to late 1800's. It's so good I make it a lot.

Carol Gordon
Topeka, Kan.

## OLD TIME COUGH CURE

Take 1 ounce each of licorice root, thoroughwort, flax seed, and slippery elm bark. Cut the elm bark and licorice root up fine. Mix with the water. Steep slowly for ten hours. Strain, and add to the syrup one pound of loaf sugar and one pint of molasses. Boil a few minutes and bottle. Take a tablespoonful four times a day.

Mary Jane Nestler
Atchison, Kan.

## ANOTHER COUGH SYRUP

Take a small handful of hops and some old field balsam and some horehound, and make a strong tea. Strain and put as much molasses as tea; boil down to about one-half. To be taken before eating and before going to bed.

Mary Jane Nestler<br>Atchison, Kan.

## A HEALING OINTMENT

Take the inside bark of sweet elder, boil to a strong infusion. Strain it, then add equal parts of beeswax and mutton tallow; say to 1/2 pint of the liquid a piece of mutton tallow and beeswax each the size of a hen's egg. Simmer until the water is out. If a softer ointment is desired, use fresh butter instead of mutton tallow. Here you have a recipe for an ointment that is invaluable as a healing remedy for sores, cuts, chilblains, and especially excellent for burns.

Mary Jane Nestler<br>Atchison, Kan.

## BEATEN BISCUITS

1 quart flour
1/2 cup butter or margarine

1/2 teaspoon salt
1/3 cup sugar

Make a stiff dough with sweet milk. Knead for smoothness. Then beat hard with a rolling pin for 15 to 20 minutes (a good job for the small boy in the family). Roll out and cut into small rounds, prick with a fork and bake in a hot oven. Good hot or cold.

### For Breakfast Rolls:
Make out rolls before going to bed. Let stand at room temperature. (If winter, cover with a wool blanket.) Bake as usual.

### For Sweet Rolls:
Roll in a sugar and cinnamon mixture, add raisins and nuts, dot with

more margarine or butter on top. Scatter remaining mixture over the top and bake.

I regret that no baking temperature or time for baking was recorded.

Betty Laverty<br>Gladstone, Mo.

## PRAIRIE PIE

2 eggs slightly beaten
1/4 cup flour
1 cup black walnuts
  (I use half black and half English)
1 cup chocolate chips
1 cup sugar
1/4 pound soft butter (oleo will work)
1 teaspoon vanilla
9-inch pie shell

Combine ingredients and pour into shell. Bake 45 minutes at 350º until set. May be frozen.

Great-Grandmother's copy called for shaved chocolate and prairie walnuts. The original copy came from Virginia to Kansas about 1860.

Carol Gordon<br>Topeka, Kan.

## HOMEMADE NOODLES

2 cups flour (measure before sifting)
1/2 teaspoon salt
1 teaspoon baking powder
2 eggs well beaten
1/2 cup cold water, added to eggs

Mix well and roll thin. Place in sun to dry, roll like a jelly roll, and slice fine. Put into boiling water and cool slowly 15 minutes. (I don't dry mine.) Simmer for 30 minutes.

(The note about not drying was written by my mom. The recipe was obtained from my Aunt Dessie Gilleand, Pawhuska, Okla. It's an old one, but still a family favorite.)

Betty Laverty<br>Gladstone, Mo.

# Rebuilding The Country
## 1866-1899

# Rebuilding the Country 1866-1899

## Saloon life was colorful in the late 1800's

The war ended. And while the South licked its wounds and rebuilt its economy, many sought their fortunes in the West. The westward expansion, well underway at the start of the war, was far from complete, and the excitement of the 1849 gold strike at Sutter's Mill in California was still fresh in many minds.

On the way, a young fortune seeker could find his own fortune depleted in a boisterous cowtown poker game. *The McCormick Gazette,* published by the McCormick Distilling Company, Weston, Mo., compiled some interesting observations from newspaper accounts of the day.

"Dodge City is dull at this time," wrote a Kansas newspaperman in 1877. "At this writing there are only 17 saloons and dance houses, 60 prostitutes, 30 gamblers, and 80 cowboys in the entire town."

The McCormick biographer called the Kansas cowtowns of that era "naughty, gay and gaudy." In Wichita, in 1874, another newsman described the scene:

"It is a motley crowd you see. Broad-brimmed and spurred Texans, farmers, keen businessmen, real estate agents, land seekers, greasers, hungry lawyers, gamblers, women with white sunbonnets and shoes of a certain pattern...the taxes are paid by the money received from the whiskey sellers, gambling halls and demi monde."

Early Kansas history records that saloons outnumbered other businesses, two to one. Unlike the modern television versions of the saloons, they were anything but plush palaces of pleasure.

In the early frontier settlements, the saloon most likely was a tent, a "soddie" or even a dugout, with planks stretched across two beer kegs. As "civilization" came, the saloons became fancier--long, narrow rooms with a bar and some tables and chairs. No decorations, no pretty girls dancing the can-can. Walls most likely were bare planks. The typical bar may have been nicer than the rest of the saloon--perhaps made of mahogany or walnut, ornately carved, with a polished glass mirror and even a crude picture of a nude female.

Fine whiskeys, not the fabled "red-eye," were stocked, and the better bars may have had some sour mash which came overland from Weston, Mo. Of course there was plenty of rotgut and snakehead, but there were fine liqueurs, the best brandies, and

## Rebuilding The Country

"The Faro Players," an illustration by W.L. Dodge.

top bartenders who could mix the latest drink popular in Kansas City or "way back on the East Coast." The beer was plentiful and cold, for there was lots of ice to chill it for the hot days on the prairie. Many of the better saloons offered music, too.

Food was on tap in some of the cowtown saloons, and sometimes the patron left himself open to a bit of fun, as the *Dodge City Times* reported on July 26, 1878:

"A good story is told of a well known citizen of this city, whose name we suppress. The story runs in this line. He went into ...'s saloon, took a seat, threw his feet on the table, and called for a glass of beer, a sandwich and some Limburger cheese, which was promptly placed on the table beside his feet. He called to ... and told him that the cheese was of no account, as he could not smell it, whereupon the proprietor replied: "Dammit, take your feet down and give the cheese a chance."

## Passing time in the 1800's meant telling "tall tales"

The telling of "tall tales" was an art form developed by the settlers, many of whom treated conversation with a higher regard than our present television-viewing population. The late Francis Williams, who wrote for the old *Discover North* magazine for several years, related this "tall tale." The tale illustrates the effect of more than one "teller" collaborating:

A group of farmers was

watching sparks fly from the forge in Gordon Clements' blacksmith shop in Platte City in the 1870's. The men were discussing dogs when David Newman of the Edgerton community began telling of Old Buck, his highly intelligent coon hound.

"Before he disappeared about a month ago, I went out every morning and hung twelve skin-stretching boards on the side of the woodshed," Newman said. "Old Buck would lay around all day with his eyes on those boards, and when we went out that night, he'd tree one coon to fit each of those boards. But I guess I played a dirty trick on that old hound. Just for devilment, I hung an old ironing board out in place of my largest stretching board, and that night when we went out, that dog headed south along Platte river and we've never seen him since. So I take it Old Buck is still looking for a pelt big enough to fit that ironing board."

Newman was concluding his story when in walked one of his neighbors. "You telling about Old Buck?" the neighbor asked.

Getting an affirmative nod, the neighbor said, "I've been meaning to tell you--I was down at Farley a couple of days ago and saw Doc Holt. Doc said he saw Old Buck sniffing at some bear tracks along the mouth of the river, so I guess that derned hound is determined not to let you down."

George Mason moved from Kentucky to Platte County, Mo. with his parents and in 1877, when 30 years old, formed a partnership known as "Mason and Flannery," and opened a large mercantile store in Platte City. Mason was credited with creating the tale of a "friend back in Kentucky" who was digging potatoes with a spade when a monstrous rattlesnake bit the wooden spade handle.

The poison, Mason declared, swelled the handle until it broke the ferrule. The farmer tossed the useless handle aside and it kept on swelling.

Finally, Mason said, when the handle swelled to seven feet thick and 163 feet long, the farmer sawed it up and used the lumber to build a tobacco barn. All went well, Mason said, until the barn was filled with tobacco.

"You know, tobacco kills snake poison," Mason said. "And when that barn was filled, the burley kept drawing out poison until all you could find of that barn was pieces of wood about the size of a toothpick."

Mason concluded by exhibiting a splinter of wood he took from his shirt pocket. "See this?" he said, holding up the splinter. "My friend gave this to me for a souvenir--it was one of the timbers that held up the barn roof."

Oscar Berry, a Confederate Army veteran and a personal friend of the infamous outlaw, Jesse James, was in his 80's when he told this tale to Ward Keith, a musician from Platte County, Mo. Keith later appeared on the Bob Burns show, and Burns told it on nationwide radio. But Berry claimed he heard it from Jesse James when the two were hiding in a straw stack on the farm of Dr. T.E. Hammond. Kansas City's Marriott Hotel, near KCI Airport, was built on this farm. Berry, who died in the early 1920's, said James received a bullet wound elsewhere and selected the hideout so Dr. Hammond could give him medical attention.

"A family of seven in Arkansas was so poor," Jesse said, "that when meat was on the menu, the mother could afford to cook only "one piece around." That night, the family expected company, and the woman cooked eight pieces of meat.

"The company didn't show up and when they were all finished eating their piece of side meat, they sat there with their mouths watering, wondering what would happen to that extra piece of meat," Jesse said.

"About that time a gust of wind blew out the coal oil lamp, and then a scream was heard. It sounded like a bobcat got caught in a steel trap. When their pappy finally got the lamp lit, the youngest brother had his fork stuck in the extra piece of meat, and six other forks were sticking in the back of little brother's hand."

## '4-horsepower ferry' served Carrollton, Mo.

The late 1800's marked a surge of riverboat activity on the Missouri River that lasted into the early 1900's. Capt. Henry K. Thomas, who was a "third generation river boat pilot," operated the Thomas ferryboat *Lillian* between Waverly and Carroll County. His grandfather had started the operation shortly after the Civil War, and it continued by Henry and his father until the Waverly bridge

The "General Harrison," the "4-horsepower ferry."

was built in 1925.

In the historical society's "His-tory of Carroll County," Thomas recalled another boat, owned by his grandfather, which plied the river in the 1880's. The *Gen. Harrison* was powered by four horses, which operated a treadmill.

"The cost of the fuel was nil," Thomas said, "because between trips, the horses, named Tom, Charley, Judy and Kate, were released from their stalls to graze near the landings. When another southbound or westward-bound covered wagon would show up to cross the Missouri River, the horses were trained so that the boat's bell would bring them aboard and into their own stalls. They would start treading, and the boat was off."

"To "full-stroke her," the pilot would call to the deckhand and the passengers to pull hard on the tails of the horses, causing them to walk faster and pull harder on the treads," Thomas said.

## First fire department in Weston, Mo. proved Murphy's Law

Back in the 1890's, a group of citizens in Weston, Mo. organized the town's first volunteer fire department. Julius Rumpel was one of the founders, and told many of the details to Dr. R.J. Felling, a contributor to *Discover North*, our original publication.

In Dr. Felling's account, he described the early-day fire-fighting group as demon-strators of "Murphy's Law," to wit: "Anything that CAN go wrong WILL go wrong."

According to the narrative, a great deal of firefighting equipment was purchased, and stored at the old Royal Brewery property (where America Bowman restaurant is now.) The main peice of equipment was a 60-gallon water tank on two wheels, "pulled by as many men who could get hold of the 100 foot rope." Ten or twelve red buckets swung from the side of the tank, and there was a pumper vehicle which resembled a railroad hand-car with a large reel of 4-inch hose that could be hooked to the pump.

The town's fire bell set on a tall tripod behind the Methodist Church, and did double duty on Sundays calling the faithful to worship.

Large cisterns, forty feet deep and twenty feet in diameter, were dug at Thomas and Market Streets where they intersected Main Street. There were also two elevated water towers on the Royal Brewery grounds.

A trial run was conducted Nov. 30, 1896--a Sunday afternoon. Crowds lined Main Street and cheered as the bell rang, and the volunteers, decked out with new red helmets, charged to a preset blaze on the river bank.

Then Murphy's Law took

Weston's fire brigade, on parade in 1896. (Weston Historical Museum photo)

effect. When the "pumper" was set up, sparks ignited the hose in several places, spraying water on the firefighters, but only a "few dribbles" came from the nozzle. The cistern cover, hardened with mud, refused to budge. The desperate crew pulled the pump across the railroad tracks and down to the river bank, threw the intake hose into the water and water was soon being played on the burning house. But no one thought to flag down a passing train, and the hose was quickly severed, bringing firefighting efforts to a halt.

The refreshment committee had done its job, however, and beer was soon at the scene, courtesy of Royal Brewery. But a dozen tin cups were left at the firehouse, and the firefighters had to use their new helmets. Julius Rumpel, the fire chief, admonished them to clean their helmets when they were finished.

Members of Weston's first fire brigade. (Weston Historical Museum photo)

## HARVESTER CAKE

(From the Harvester Restaurant, Chillicothe, Ohio. This restaurant has been in existence since the late 1800's)

1 1/2 cups firmly packed brown sugar
1 1/2 cups granulated sugar
3 eggs
1 1/2 sticks (3/4 cup) butter, melted
3 teaspoons ground cinnamon
1 1/2 teaspoons baking soda
3/4 teaspoon salt
3/4 teaspoon baking powder
2 1/4 cups flour
1 1/2 cups quick-cooking rolled oats, cooked in 2 cups boiling water.

In large bowl, mix all ingredients in order given. Pour into greased 13x9 inch pan and bake at 350° about 40 minutes or until firm. Do not overbake. While still warm, spread with topping and broil until bubbly and brown. Cut into squares. Makes 12 servings.

### TOPPING

1 1/2 cup firmily packed brown sugar
1 1/2 sticks (3/4 cup) butter, melted
6 ounces evaporated milk
3 cups grated coconut
1 1/2 teaspoon vanilla

Blend. This is a favorite recipe that has been in our family for many years.

Louise M. Bell<br>Utica, Ohio

## MOLASSES CANDY

1 cupful of New Orleans molasses
1 tablespoon of vinegar
Piece of butter the size of an egg

Boil, but do not stir, until it hardens when dropped in cold water. Watch it that it does not burn. When it becomes hard and brittle, stir in a teaspoonful of soda and beat well; pour into buttered pan. When cool, pull until yellow, using butter on your hands so that the candy will not stick.

Eileen Filley<br>Cameron, Mo.

## TO RESTORE GILT FRAMES

With a soft brush carefully free the frames of every particle of dust, then cover with the following mixture: the white of an egg and 1/2 ounce of chloride of potassa. Apply with a soft brush.

The *Scientific American* says if you desire to cleanse gilt frames without tarnishing them, wash them in beer.

Eileen Filley<br>Cameron, Mo.

## CEMENT FOR CRACKS IN STOVES AND STOVEPIPES

A good cement may be made of wood ashes and salt, in equal parts. Make a paste with cold water and fill the cracks when the stove is cool. It soon hardens.

Eileen Filley<br>Cameron, Mo.

## HOW TO CLEAN MICA IN STOVES

Never attempt to clean the mica with water and soap. It will cause it to scale at once. Dip a soft cloth in clear vinegar and rub the mica over quickly, not forgetting the corners. It will stay clean for a long time.

Eileen Filley<br>Cameron, Mo.

## HOW TO CLEAN VINEGAR CRUETS

Shake crushed eggshells and a little water vigorously in a vinegar cruet and it will remove that cloudy look which the bottle often takes on.

Eileen Filley<br>Cameron, Mo.

## HOT SWEET MUSTARD

3/4 cup Coleman's dry mustard  
2 well-beaten eggs  
1 cup vinegar  
1 cup sugar

Put dry mustard in the vinegar and allow to stand 2 hours. Mix in the eggs and sugar. Cook in double boiler, stirring constantly until thick.

Mrs. George Rittman  
Weston, Mo.

## VINEGAR PIE

Line pie tin with good crust; take 3 tablespoons of flour, one teacupful of sugar and mix well. Now add 3/4 cupful of good vinegar and 1/4 cup of water; flavor with nutmeg and cover with strips of crust. Lay on bits of butter and bake in moderate oven.

Eileen Filley  
Cameron, Mo.

## VINEGAR ICING

BLEND:  
1 tablespoon sugar, 1 cup powdered sugar

MIX WELL:  
2 well-beaten egg whites, 1 tablespoon vinegar, 1/2 teaspoon vanilla

If this is not stiff enough to spread, add more powdered sugar. This is an old-timer, especially good on apple cake.

Mary Jo Ben  
Kansas City, Mo.

## SHROVE TUESDAY PANCAKES

Beat the yolks of 6 eggs with 4 cups of whole milk. Add one teaspoon of salt and 2 1/2 cups of flour. Make into a batter.

Beat whites of 6 eggs stiff and mix in batter. Pour on a hot greased griddle and sprinkle with currants.
Serve with brown sugar and fresh lemon juice.

"We always had these pancakes for the evening meal on Shrove Tuesday when I was growing up--yummy! You lay a pancake on your plate, put brown sugar on it, then squeeze lemon juice over it. My father said he had them in his house as a boy. He was born in 1880."

Lulua Fillmore<br>Independence, Mo.

## CASTOR OIL COOKIES

1 cup sugar
1/2 teaspooon salt
1 teaspoon soda
2 teaspoons ginger

1 cup molasses
1 cup milk
1/2 cup castor oil
Flour

Mix all ingredients well. Add enough flour to make a dough to make a firm ball that can be rolled. Cut out and bake in 375º oven.

Two cookies are one dose of castor oil. "This recipe was handed down for more than 80 years."

June Medina<br>Baldwin City, Kan.

## GUM DROP COOKIES

1 cup shortening
1 cup white sugar
1 teaspoon baking powder
1 teaspoon salt
1 cup shredded coconut
2 cups flour

1 cup brown sugar
2 eggs
1 teaspoon soda
2 cups quick oats
1 cup cut gum drops

Mix, roll in balls, press flat and bake on ungreased pan at 375º for 10 minutes.

Mrs. George Rittman<br>Weston, Mo.

## CHICKEN PIE

3 cups diced cooked chicken
1/2 cup sweet milk
5 tablespoons flour
8 1/2 ounce can peas
8 ounce sliced mushrooms
1 cup mushroom soup
1 cup chopped celery

3 eggs
4 tablespoons oleomargarine
1 1/4 teaspoon each salt, pepper and meat flavoring
2 ounces pimentos
1 cup chicken broth

Cook celery in pea juice and chicken broth until tender. Beat eggs and mix with milk, flour and margarine. Add rest of ingredients and put in large cake pan. Top with the following pastry:

1 cup flour
1/2 cup shortening
Pinch of salt

1 egg white
1 tablespoon white syrup

Mix this until it is a coarse mixture and add 1/2 cup of grated cheese. Roll out and top the chicken mixture. Bake at 400º until brown.

Mrs. George Rittman
Weston, Mo.

## GRANDMOTHER OSBURN'S BREAD & BUTTER PICKLES

1 dozen cucumbers

1/2 dozen onions

Slice with peeling on. Put in separate jars in salt water for one hour. Rinse in cold water and drain.

1 pint vinegar
1 teaspoon pepper
1 teaspoon ginger
1 teaspoon cinnamon

1 cup sugar
1 teaspooon celery seed
1 teaspoon mustard
1 teaspoon tumeric

Cook over a slow fire. This recipe is about a hundred years old.

Beverly Rizzo
Kansas City, Mo.

## SHIPWRECK CASSEROLE

In a greased casserole, layer 1 1/2 pounds hamburger, seasoned to your liking. Then add:

onions, sliced
can red beans

6 potatoes, sliced
1/2 cup uncooked rice

Top with 1 can tomato soup and enough water to cover. Bake covered for 1 hour at 350°.

Mrs. George Rittman
Weston, Mo.

## BOILED RAISIN CAKE

1 cup raisins
1 cup water
1/2 cup lard
1/2 teaspoon salt

1 cup sugar
1 teaspoon ground cinnamon
1/2 teaspoon ground cloves
1/2 teaspoon nutmeg

Boil together for 20 minutes and cool after boiling. After cooling, add:

1 egg beaten

1 teaspoon baking soda

2 cups flour (scant) bleached
  or unbleached
Nuts may be added, or other fruits
  for fruitcake

Bake at 350°. Pan size depends on amount of fruit used. This recipe has been in our family for at least three generations.

Beverly Rizzo
Kansas City, Mo.

## BOILED APPLE DUMPLINGS

Make rich biscuit dough such as soda or baking powder biscuits, only adding more shortening. Take small piece of dough, roll as thin as pie dough for pie. Cut in pieces large enough to cover an apple. Put in middle of square pared and cored apple. Sprinkle with spoonful of sugar and couple pinches of cinnamon. Turn corners over apple and lap tight.

Put in square cloth well floured dipped in hot water. Tie good bu leave enough room for dumpling to swell. Put in pot of boiling water and boil 3/4 hour. Serve with cream.

This was originally published in a "White House Cookbook" in 1887. It's been used by my family ever since.

F. Maxine Adam
Fulton, Mo

## PLUM PUDDING

MIX TOGETHER:
1 cup suet (cut fine)
1 cup sorghum
1/2 cup brown sugar
1/2 cup buttermilk

SIFT TOGETHER:
2 cups flour
1 teaspoon soda
1 teaspoon baking powder
1 teaspoon cinnamon

OTHER INGREDIENTS:
1 cup raisins
1 cup black walnuts

Add the flour mixture to the suet mixture, blending well. Then the raisins and nuts (and candied fruit if you like). Pour into a greased baking dish and bake at 250º for three hours.

"Grandma served it warm with this hard sauce:"

1/3 cup butter or margarine
1 cup powdered sugar
1 tablespoon lemon juice

Cream the margarine, add sugar gradually, then lemon juice. Store in the refrigerator.

Edith Snide
Kansas City, Mo

## CUSTARD

*(From an 1867 recipe book printed in German and English)*

Whip with a rod, eight whole eggs, a pint and a half of cold milk, a few drops of lemon juice, half an ounce of sugar and a very little salt.

Then place the whole several times through a hair-sieve, and stir it on a slow fire till it begins to curdle.

Take it at once from the fire, and stir for a minute longer, and then put it to cool in proper custard forms.

Now make a cream or vanilla sauce, and let it get cold. In summer put it on ice.

Frances Uttinger<br>Kansas City, Kan.

## GRANDMA OPAL'S ICE BOX ROLLS

Mix Together:

| | |
|---|---|
| 1 cup warm water | 1/2 cup sugar |
| 1/3 cup shortening | 3 teaspoons salt |

PART 2:

| | |
|---|---|
| 1 cup water | 2 packages yeast |

Beat 2 eggs well, add Part 2 and beat. Then add to first mixture and beat well.

Add 7 cups flour, 1 cup at a time. Knead a stroke or two. Put in a greased bowl, let rise one hour at room temperature.

Wrap with plastic wrap and keep in the refrigerator. As you need, work down dough and pan out what is needed.

Let rise and bake at 350º in a lightly greased pan until done.

This recipe is very old--and very good.

Kathy Osborne<br>Sedalia, Mo.

## VANILLA CUSTARD Whipt on the Fire
*(Also from the 1867 German-English cookbook)*

Stir a spoonful of rice flour with the yolks of eight eggs and three half pints or a quart of milk or sweet cream, till smooth; put in half a stick of vanilla, and now whip it, over a gentle fire, till it has thickened a little, and looks like a nice thin batter.

Empty it into another vessel, season it with five ounces of sugar and a little salt, and stir it slowly till quite cold, in summer, on ice. In dishing it, add to it a pint of whipt cream, stirring it in gently, then send it to the table in a puff-paste crust or in a deep dish.

Frances Uttinger<br>Kansas City, Kan.

## LEBKUCHEN COOKIES

1 pint honey
1 egg (beaten)
2 teaspoons soda  in 1/2 cups sour milk
1/2 cup nuts
1/2 cup candied fruit
1 tablespoon butter

1 cup molasses
1 1/4 pounds brown sugar
2 teaspoons cinnamon
1 1/4 teaspoons cloves
1/2 teaspoon salt

Mix all ingredients, and heat till mixture blends and softens (not too hot). Cool and add nine cups of flour. Mix and put in icebox overnight. Use plenty of flour to roll and cut in strips.   Bake at 350º.   Remove immediately from pan or will stick. May glace while warm. Makes a huge batch and they keep indefinitely.

This recipe came from a lady whose mother brought the original from Germany. It was in German and she had it translated.

Beverly Rizzo<br>Kansas City, Mo.

## GINGERBREAD

2 eggs
1 cup sugar
1 cup molasses
1/2 cup shortening
1 teaspoon ginger and cinnamon

1 cup boiling water
1 teaspoon soda
1 teaspoon baking powder
2 1/2 cups flour
1/2 teaspoon cloves

Cream sugar and shortening, add beaten eggs, molasses, and boiling water. Sift dry ingredients and add to liquid. Bake in moderate oven.

This is another recipe that has been in our family for more than three generations.

Beverly Rizzo<br>Kansas City, Mo.

## OATMEAL CAKE

In a large mixing bowl:

1 cup quick oatmeal
1 1/4 cup boiling water

1 stick oleo

Cover and set for 20 minutes. Then add:

1 cup brown sugar
2 eggs

1 cup white sugar

Sift and add:

1 1/3 cup flour
1 teaspoon cinnamon
1/2 teaspoon nutmeg

1 teaspoon soda
1/2 teaspoon salt

Mix together and bake for 30 to 35 minutes in 350° oven.

### Topping for Oatmeal Cake

6 teaspoons soft butter or margarine
1/3 cup sugar
1/4 cup cream or evaporated milk

1 cup large nut pieces
1/2 teaspoon vanilla
1 cup coconut

Spread over cake and place under broiler until brown.

Mrs. George Rittman
Weston, Mo.

## MOLASSES PIE

2 teacups of molasses
3 eggs
1 lemon, grated and squeezed

1 teacup sugar
1 tablespoon melted butter
nutmeg

Beat well, bake in pastry. This is another family recipe dating back before the turn of the century.

F. Maxine Adams
Kansas City, Mo.

## PEPPER RELISH

24 green peppers           24 red peppers
12 large onions

Grind through the food chopper (today we would use a food processor); cover all with boiling water and let stand ten minutes. Drain well. Add:

1 quart vinegar           4 cups sugar
3 tablespoons salt

Boil for 20 minutes. Place in sterile jars and seal. Great in tuna salad and in salad dressing for sandwich spread..

Mrs. George Rittman
Weston, Mo.

## SUCCOTASH

Take a pint of fresh shelled lima beans. Put in pot with cold water, rather more than will cover them. Scrape the kernels from 12 small ears of fresh sweet corn. Put cobs in with beans, boiling 1/2 to 3/4 hour. Now take cobs out, put in scraped corn. Boil again 15 minutes, then season with salt and pepper, piece of butter the size of an egg and 1/2 cup cream. Serve hot. An old recipe from 1880s.

F. Maxine Adams
Fulton, Mo.

## LEMON BUTTER

1 1/2 cups sugar         3 egg whites
1 egg yolk                 1 1/2 cup butter
Yellow and juice of 2 medium size lemons. Yellow means grated rinds.

Cook 20 minutes in double boiler and pour in glasses as jelly.

Mrs. Charles Fryer
Kansas City, Mo.

## BUCKWHEAT PANCAKES

"I can trace this recipe back to my great-grandmother Neal, who served them as a hearty winter breakfast to her family. My grandmother kept a "starter" going all winter, as did my mother when I was a child. My family were all farming people, and a big breakfast was a necessity for the hard physical labor entailed in farming early in the century.

Nowadays, those of us in our family who like them serve the pancakes as a special supper with homemade brown sugar syrup and ham or sausage, and as a Christmas Eve treat."

### Starter

1 package dry yeast            1 tablespoon sugar

Dissolve the above ingredients in 2 tablespoons of warm water and add:

1 cup buckwheat flour           2 cups white flour
2 cups sour milk or buttermilk

Let stand in warm place to rise. I mix it before noon to serve for supper, or can be mixed to rise overnight. Thirty minutes before meal-time, add 1 teaspoon salt, 1 tablespoon sorghum or 1 tablespoon brown sugar syrup, and 1 teaspoon soda dissolved in 2/3 cup boiling water. Stir into batter. Add more boiling water if necessary to make the batter the consistency you like.

Bake on hot greased griddle and serve with butter and syrup. I use old-fashioned cast-iron griddles, but modern ones, well-greased, will work OK. Leftover batter can be kept in refrigerator and used as starter, if used within a week.

Florence Flanary
Burlington Junction, Mo.

## GINGER BREAD

1/2 cup sugar
1 egg
2 1/2 cups flour
1/2 teaspoon ginger
1/2 teaspoon salt

1/2 cup shortening
1 cup molasses
1 1/2 teaspoon soda
1/2 teaspoon cloves
1 cup hot water

Cream shortening and sugar. Add beaten egg, molasses, then the dry ingredients sifted together. Add hot water last. Batter will be soft. Bake in greased shallow pan about 35 minutes in moderate oven.

LaVon David
Prairie du Chien, Wis.

## GRANDMA BRYAN'S LEPPE COOKIES

1/2 gallon molasses
1/2 cup cinnamon
1 tablespoon cloves
2 cups lard
4 cups buttermilk
1 box raisins (large)

1 to 2 packages of dates

1 cup sugar
1 tablespoon allspice
1 1/2 tablespoons nutmeg
1/2 cup soda
4 cups nuts
4-6 cups fruit (candied cherries and pineapple preferred, not citron)
Flour to make stiff dough

Grind fruit, raisins and dates with small blade of food grinder. Use next largest blade for cherries and coarsest blade for nuts. Mix all ingredients in clean dishpan or large kettle. Cover and chill overnight. Let dough come to room temperature and roll out on lightly floured surface. Cut into shapes with cookie cutters or use sharp knife. Bake on ungreased cookie sheet at 350º (do not overbake). These are so good iced after cooling with the following icing:

2 cups brown sugar
Butter (size of walnut)

1/2 cup cream

Cook all to soft ball stage. Beat till it loses gloss. Add 1 teaspoon vanilla and ice cookies. I freeze dough in rolls, thaw slightly, slice with sharp knife and bake as needed.

Sharon Mellor
Liberty, Mo.

## MOM MELLOR'S LEBKUCHEN (LEPPE) COOKIES

1 quart molasses
1 1/2 pound dates
3/4 pound butter (1 1/2 cup)
2 tablespoons cinnamon
2 tablespoons cloves
3 pints buttermilk
3 pounds raisins
5 pounds flour (or more) to stiffen dough

1 quart brown sugar
3/4 pound lard (1 1/2 cup)
1 quart white sugar
2 tablespoons allspice
2 tablespoons ginger
1/2 cup soda
1 quart nuts (at least)

Mix all ingredients and let stand overnight. Roll out and cut into squares or shapes. Bake at 350°. Can be iced and decorated when cool. Improve with age and freeze well.

Sharon Mellor
Liberty, Mo.

## APPLE CRISP

5 or 6 apples, sliced
1 teaspoon cinnamon

3/4 cup sugar

Place apples in baking dish, and sprinkle on sugar and cinnamon. Then mix the following ingredients:

3/4 cup oatmeal
1/2 cup flour

3/4 cup brown sugar
1/2 cup margarine

Cut margarine into flour until margarine is coarse in size. Mix together with oatmeal and brown sugar. Blend well. Spread over apples and bake at 350° about 45 minutes or until golden brown.

Marilyn Holliman
Farragut, Iowa

## CHICKEN AND 'SLICKERS'

"My mother-in-law and her mother before her used this recipe, and it's one of my favorites. My mother-in-law raised nine children (the oldest is now 87 years old--1988) back when times were really hard. She baked biscuits every morning and made four loaves of bread every other day. In the winter she cooked on a wood stove and in the summer on a kerosene stove."

Stew one stewing chicken until well done. Be sure to use enough water so that you have plenty of chicken broth. When done, remove chicken and cut into pieces. Add cut-up chicken to broth.

### Slickers

2 cups flour
1 teaspoon salt
2/3 cup milk

3 teaspoon baking powder
1/3 cup shortening

Cut shortening into flour, baking powder and salt mixture until it is like coarse cornmeal. Add milk, stirring it in quickly to form a soft dough. Turn dough out onto a lightly floured surface. Knead about one minute. Roll out about 1/2 inch thick, cut dough into strips about 1 1/2 to 2 inches wide and 5 inches long. Drop into boiling chicken broth and cook over low heat 10 minutes. Cover and cook 10 more minutes.

For a real treat, slice two or three medium sized potatoes thin and drop the slices into the broth with the slickers.

Mrs. Dewey Blackburn
Hamilton, Mo.

## BANANA-CARROT NUTBREAD

2 cups flour
1/2 teaspoon salt
1 cup mashed ripe bananas
3/4 cup vegetable oill
1 cup finely grated pared carrots

1 teaspoon baking soda
1/2 teaspoon cinnamon
1 cup sugar
2 eggs
1/2 cup chopped pecans

Sift together flour, baking soda, salt and cinnamon. Set aside. Combine bananas, sugar, oil and eggs. Beat with electric mixer at medium speed for two minutes. Stir in dry ingredients. Fold in carrots and pecans. Put in greased and floured 9x5x3 loaf pan. Bake at 350º for 55 minutes or until a wooden toothpick inserted in center comes out clean.

Marilyn Holliman
Farragut, Ia.

## DATE PUDDING

This is one of my favorite recipes, and is one my grandmother made. When I was a little girl it was a Christmas treat for us. Grandmother baked this in a crock and never had any kind of oven except a kerosene oven.

### Batter

1 cup sugar
1 cup nuts (black walnuts)
1/2 cup milk

1 cup dates, cut up
1 cup flour
2 teaspoons baking powder

Mix all ingredients well and pour into a greased 9x9 baking pan.

### Sauce

2 cups boiling water
1 cup brown sugar

Butter the size of an egg

Stir until well dissolved and pour over batter. Bake in a 350° oven for 30 minutes.

Mrs. Dewey Blackburn
Hamilton, Mo.

## CINNAMON WAFFLES (Aimmet Waffles)

1/2 pound butter
3 eggs
2 teaspoons cinnamon

1 cup sugar
Flour

Cream the butter and sugar, beat in the eggs one at a time, add cinnamon, work in enough flour to make a soft dough. Form into small balls. Place several in a hot waffle iron suitably spaced. Press down top and bake.

This is an old recipe from Germany.

Mary Jane Nestler
Atchison, Kan.

## CRANBERRY PUDDING

1/3 cup flour
2 teaspoons soda
2 cups raw cranberries

1/2 cup molasses
1/3 cup hot water
1/2 cup nuts

Dissolve soda in molasses. Add 1/3 cup hot water, cranberries, nutmeats, and flour. Put in greased pan. Steam two hours.

## SAUCE

1/2 cup butter
1 tablespoon flour

1 cup sugar
1 cup cream

Make like a white sauce. Use vanilla or rum for flavoring.

LaVon David
Prairie du Chien, Wis.

## SUET PUDDING

1 cup suet, chopped or ground fine
1 cup milk
1 cup raisins mixed with 1/4 cup flour
1 teaspoon cinnamon
2 cups flour

1 egg
1 cup molasses
1/2 teaspoon allspice
1 1/4 teaspoons soda

Steam 3 hours.

LaVon David
Prairie du Chien, Wis.

## GREAT-AUNT ROSA'S WATERMELON RIND PRESERVES

2 tablespoons crumbled lime
2 tablespoons ground ginger
2 lemons

2 pounds watermelon rind
2 pounds sugar

Cut away all the green skin and red watermelon from the rind. Thick rind is the best. Cut the rind into small pieces or strips. Dissolve lime in 8 cups cold water and pour over the rind. Then soak about 3 hours. Make sure you rinse the rind really good after it's been soaked and then let it stand another half hour in clean water. Drain all the water off and then sprinkle the ginger over the rind and cover again with water.

Boil until it's tender when you prick with a fork. Drain again. Put in sugar and juice of one lemon to 7 cups water and boil 5 minutes. Cool real good and add the rind to the syrup. Boil another 30 minutes and then add the other lemon, sliced thinly. Cook until the rind is clear. Pack into hot jars and process 20 minutes in simmering hot-water bath.

Norma Rouse Middleton<br>
Pleasant Hill, Mo.

## NORMA'S BRANDIED PEACHES

6 fresh peaches, perfect as possible
2 cups sugar
brandy
3 cups water
2 tablespoons good quality

Dip perfect peaches quickly in hot water and peel. If the peaches have thin skins, you may prefer not to peel them. Rub off the fuzz with a clean cloth and prick each peach twice with a fork.

For each six peaches, boil three cups of sugar 10 minutes. Cook the peaches in the syrup until tender when tried with a toothpick (about 5 minutes). Pack into a pint jar. Add 2 tablespoons of brandy and fill jar with syrup. Seal and store at least a month before using. Makes a neat Christmas gift.

Norma Rouse Middleton<br>
Pleasant Hill, Mo.

## BLACKBERRY CORDIAL

1 quart blackberry juice
1/2 whole nutmeg
1 tablespoon allspice
1 muslin bag
1 pound white sugar
1 tablespoon cinnamon
1 pint best quality brandy
1/2 tablespoon cloves

Place nutmeg, cinnamon, allspice and cloves in a muslin bag. Boil the blackberry juice with sugar and spices 15 minutes. Skim well and add brandy. Let cool enough to bottle. Take out spice bag, bottle and seal with wax. Will keep for years and improves with age.

This recipe was taken from a notebook published by the Pleasant Hill

Presbyterian Church, 1893. It was the recipe of Mrs. F.W. Little, whose husband owned the Pleasant Hill Royal Extract Mfg. Co.)

Norma Rouse Middleton<br>
Pleasant Hill, Mo.

## DARK RED CAKE

3/4 cup cocoa  
1/2 cup sweet milk

1 cup brown sugar

Cook in double boiler until creamy, then cool. Mix with the following:

3 eggs, beaten  
1/2 cup butter--beat with eggs & sugar  
1/2 teaspoon vanilla  
2 1/2 cups flour

1 cup white sugar  
1/2 cup sweet milk  
2 teaspoons baking powder

Makes a thin batter. Bake in loaf or layer in moderate oven--350º. Sour milk can be used instead of the sweet milk (we had more sour milk in the early days, and it could be sweetened with soda, but I don't remember how much).

This was our "family cake recipe" I remember as a child. It is probably more than 100 years old.

Mrs. Charles Fryer<br>
Kansas City, Mo.

## SAUCEPAN OATMEAL COOKIES

"This was my grandmother's recipe. She made these during the summer in the days before air conditioning. Since the oven was not required, it did not make the kitchen so hot."

2 cups sugar  
1/2 cup milk  
2 cups oatmeal  
1 tablespoon corn syrup

1/4 cup cocoa  
1/2 cup margarine  
1/4 cup peanut butter

In heavy saucepan, combine sugar and cocoa. Stir in milk and add margarine. Bring to boil, stirring constantly, and boil vigorously for 3 minutes. Stir in oatmeal, syrup and peanut butter. Return to boil and remove from heat. Stir until slightly thickened. Drop by teaspoonfuls on waxed paper.

Marilyn Holliman<br>Farragut, Ia.

## RHUBARB CAKE

1/2 cup margarine
1 egg
1 teaspoon soda
1 cup sour milk

1 1/2 cups brown sugar
2 cups flour
2 cups chopped rhubarb
1/2 cup white sugar

Mix and sprinkle 1 teaspoon of cinnamon on top. Bake at 350° until toothpick comes out clean.

LaVon David<br>Prairie du Chien, Wis.

## PUMPKIN COOKIES

2 cups flour
1 teaspoon baking soda
1 teaspoon cinnamon
1 cup butter or margarine, softened
1 cup granulated sugar
1 cup pumpkin

1 cup quick or old fashioned oatmeal
1 teaspoon vanilla
1/2 teaspoon salt
1 cup firmly packed brown sugar
1 egg, slightly beaten
1 cup semi-sweet chocolate morsels

Bake at 350° until lightly browned. May be iced if desired.

LaVon David<br>Prairie du Chien, Wis.

Recipes from "homesteading days" are always interesting, and they're getting rare, as we modify them for modern use. The following are "homesteader" types from Luise Dyche, St. Joseph, who wrote: "My grandparents homesteaded in Kansas. I can remember my grandfather, T.S. Browning, telling about a sod house he built near Sparks, Kansas. In a handwritten little book of recipes ("receets," my grandmother called them) is a recipe for dried corn. In the days when Ball jars and freezers weren't available, one had to preserve vegetables and fruits in other ways. There wasn't much space in sod houses or the first frame houses so winter vegetables had to be dried or buried under ground. This recipe for dried corn was dated April 8, 1887 and written by my grandmother, Clara E. Browning:

## DRIED CORN

8 pints of corn cut off cob and scraped
1/4 pint salt
1/2 pint sugar
1 pint cream and milk

Cook 20 minutes after it begins to boil, stirring constantly. Dry on platters or large plates. Stir second day, if not sooner. Takes three days of good hot sun.

Luise Dyche
St. Joseph, Mo.

"Some of the old cookbooks were handwritten in little memo books (like the spiral books we have now). Many measurements were given by the bowl, handful, etc. One of my old books--published by various businesses in Holton, Kansas--even states that 35 cents worth of steak will serve six!

"This second recipe was my mother's. In my youth we didn't have the variety of food we have now but nothing could top a good slice of homemade bread--especially if it was still warm--with homemade butter and some of this good pear honey:"

## PEAR HONEY

Peel the pears and grind them raw. Use one gallon of raw ground pears. Add about one cup of water. Also add one 16-ounce can of crushed pineapple. Cook for one hour. Then add sugar to suit your taste, stirring frequently. Cook until real thick, and can hot.

Luise Dyche<br>
St. Joseph, Mo.

## Compton Recipes from Old Clay County, Missouri

The following recipes are from Compton family members, whose family homesteaded and built Oakridge Farm. The farm and original cabin, built in the early 1800's, was the site of Sandy's Oakridge Manor, for many years a popular tea house, and later converted to Stroud's Restaurant.

The recipes were furnished by Elizabeth Lendle, Kansas City, Mo., who got them when she worked at Oakridge Farm.

## BEATEN BISCUITS

1 quart flour
1 teaspoon salt
3/4 cup lard

2 teaspoons sugar
1 teaspoon baking powder

Make into a stiff dough with ice water and milk, half and half. Work on kneader or beat with a mallet until smooth and glossy. Roll out, cut, prick with a fork and bake slowly.

Used by Mrs. Rosa Fugitt (Compton) in the 1890's.

## LACE COOKIES

1 cup molasses
1 cup butter
1 teaspoon baking powder

1 cup sugar
2 cups flour
1 teaspoon soda

1/8 teaspoon salt

Scantly heat to boiling point numbers 1, 2 and 3. Boil one minute. Remove from fire and add 4, 5 and 6 (sifted together). Stir well. set in pan of hot water to keep from hardening. Drop 1/4 teaspoon of batter three inches apart on butterd sheet. Bake in slow oven until brown. Cool slightly. Lift off carefully with spatula.

This recipe was used by Mrs. Elizabeth Schwitzgebel, a niece of the Comptons, in the 1930's.

## CHOCOLATE CAKE

2 cups flour
1/2 cup butter
2 eggs

1 cup sugar
1/2 cup sweet milk
3/4 teaspoons soda,
    dissolve in hot water

COOK THIS:

1/4 pound chocolate
1 cup sweet milk

1 cup sugar
1 egg

Boil until smooth, then add 1 tablespoon vanilla. When cool, stir. Frost on top. This recipe was used by Mary Winn (Compton).

## HONEY CURED BACON

To 100 pounds of bacon, add:
2 pounds of sugar

10 pounds of salt
1 pound of saltpetre

Rub the salt on well, then the sugar, then sprinkle on the saltpetre. Pack in a tight box to exclude the air as much as possible. Shift the meat every seven days, adding more salt.

Always place the juice on top at the bottom, and so on, each time it is shifted. Do that for four or five weeks, or until it is well salted. Then smoke and sack as you do hams. Always use paper bags as it excludes air.

This recipe was furnished by Miss Emma Compton, from the 1890's.

## TO COOK LARD:

Cook the lard until partly done or until the cracklins will settle to the bottom. Remove from fire. Put in a large handful of salt, or about a pint to the kettle. When settled, dip out, strain and put back on fire. Keep it till it gets to 200 degrees.

This was used by A.L. Dold in the 1890's.

## MOLASSES HAND CREAM (TAFFY)

3 pounds boiling sugar

Pour slowly:
4 tablespoons black molasses,
1 tablespoon beaten butter

Boil to the third degree of boiling (third degree of boiling is to rock like an egg shell). When cool enough to remove to the slab without ruining, flavor and pull by hand or hook fastened to the wall, until drops off in strings. Then shape in a nice band and cut with large shears, any shape you wish.

This recipe was used by Emma Compton, and was dated Dec. 23, 1891.

## HULLED PECANS

1 cup sugar                 1/2 cup cold water

Boil three minutes, then add pecans. Boil for 10 minutes. Remove from fire. Stir until it creams, spread on buttered paper.

This recipe was used by Mrs. Oldham, and dated March 29, 1899.

## LEMON SYRUP

10 lemons                   1 quart water
2 quarts sugar              Green color

Add a small amount of flour stirred into cold water, enough flour to make rich but not but not thick and 1/4 teaspoon salt. Before this put through a meat grinder enough mint to make 1/2 cup of fresh mint and steep in cold water. Add juice to cold sugar syrup and enough mint juice to suit taste. Put in ice box and when ready to serve, put 2 tablespoons of water, shaved ice and enough mint to taste.

This recipe was used by Miss Emma Compton in the 1890's.

## GRAHAM CRACKER COOKIES

Graham crackers
   (to cover large jelly roll pan)
1/2 cup sugar

1 stick butter or oleo
1/2 cup sugar
Nuts (chopped)

Break crackers into sections and cover cookie sheet. Bring sugar and butter to boil in small pan. Boil 2 minutes. Pour over crackers. Sprinkle with nuts. Bake 12 minutes at 325º. (Don't let them brown too much.) Remove to rack to cool.

Sharon Mellor
Liberty, Mo.

## CRANBERRY-WALNUT RELISH

1 bag cranberries
1/2 teaspoon cinnamon
3 tablespoons lemon juice

1 1/2 cups sugar
1 cup tart orange marmalade
1 cup broken English walnuts

Wash and drain berries. Put in shallow baking dish (9x13). Stir in marmalade, sugar, cinnamon and lemon juice. Cover tightly with lid or foil. Bake at 350º for 1 hour. Put walnuts in same oven to toast last 10 minutes of baking time. Stir in walnuts and chill. Very good with turkey or poultry. Makes about 4 cups and keeps indefinitely. This is a traditional Christmas and Thanksgiving dish.

Sharon Mellor
Liberty, Mo.

# The 20th Century Begins
## 1866-1929

# The 20th Century Begins 1900-1929

## Halcyon days of the great show boats

Although the railroads were kings of transportation, the romantic appeal of the steam powered show boat was strong on the river towns at the turn of the century. Many of the big freighters had succumbed to sandbars and snags, and others had burned at their moorings or had their boilers blown apart in fiery river disasters. The ones that survived, however, meant "showtime" to riverfront crowds.

Mrs. Lutie Gordon Jordan, writing for the Carroll County Historical Society, recalled the show boats:

"Back around 1910 to 1920 people in river towns like Waverly looked forward to the coming of the show boat season," she wrote. "The two best known and remembered in this section of the river were *The Princess* and *The Wonderland*."

"Norman Thom, who owned *The Princess*, was not only captain of the boat, but leading man and director of the show. He was born in Greenup, Kentucky in 1883, and was a handsome, debonair man. Back in those show boat days, he was called the "John Drew of the River." His wife, Grace, was the leading lady and the calliope player. Their daughter, Norma Beth, like Magnolia in the stage version, played the juvenile roles. The grandmother was in charge of the boat's kitchen, or galley, in river parlance."

H.K. Thomas of Waverly, retired river captain, recounted some interesting incidents as a young man back in his early river days as a pilot one season on the ferry boat that towed *The Princess* into the Waverly dock. Not only was he the pilot but on

One of the showboats of the era, the "Jos. Kinney."

occasion, when needed, he could double in the orchestra.

In 1928 Norman Thom dismantled *The Princess,* and the producers of the screen version of Edna Ferber's well known book, "Show Boat," employed the entire cast and even used the calliope for an appearance in New York as an introduction for the opening show.

*The Wonderland,* also well remembered in this section of the river, was built in 1906 at a cost, it was said, of $40,000, and sank at Belleville, West Virginia, in 1918. It had a seating capacity of 900. There are still a few in the river towns who can recall the show boat dropping anchor and tying up at the foot of Waverly's "depot hill" for the evening show. "Waverly was always known as a good show town by the troupers in those days," Mrs. Jordan wrote. "And the surrounding towns and countryside would pour in for the unique entertainment of the show boats."

Mildred F. Burns, writing in the old *Discover North* in the early 1980's, remembered her childhood in several articles. The daughter of a farm family in Northwest Missouri, her life was typical of most rural Americans in the early 1900's. Life moved at a slower pace then, but her family was close and self-sufficient.

Her stories always left the reader with a good feeling, and a sense of admiration for those who lived in that simpler time. The next ones were my favorites:

## Eating well on the farm ...in 1910

Our ideas of good nutrition have changed considerably since 1910. And the variety of foods that are available to us has increased sharply. But farm families in the early days, although they consumed fattening foods, frequently "burned it off" with rigorous exercise that would make our modern "aerobics" seem tame. Mildred S. Burns remembers how it was:

"We were well-fed as we grew up on our little farm in northwest Missouri from about 1910 until 1930. My three brothers, my parents and I ate lots of meat, bread and sweets. We never discarded the fat on our meat. Not one of us was overweight which had to be the result of our engaging in lots of physical activity.

"I was awakened each morning by the noises Papa made shaking down the ashes in the big black range in the kitchen. He was getting a fire started for Mama to cook breakfast. Our men did hard work out-of-doors so our first meal of the day was substantial. We started with ham or bacon, gravy and biscuits. Papa helped cook breakfast. He took special interest in the gravy. The picture he made I'll never forget. He stirred so vigorously that his whole body was involved in the movement.

"There was always fresh country butter on our table along with honey, sorghum and preserves. Each of us topped off his breakfast with an enormous dish of hot oatmeal served with sugar and plenty of thick cream from our Jersey cows. My brothers and I could drink all the milk we wanted. Our parents had coffee, but it was not allowed for children.

"We produced much of our food. We always had lots of pork. It was Papa's habit to have eight or ten large hogs butchered every winter. Our storehouse hung full of hams from season to the next. The old hams --those left hanging into the second year--were especially prized. We did not smoke the meat. We learned to wrap it in a mixture of brown sugar, salt and pepper. We injected an artificial smoke flavoring with a long needle. The sides of bacon were left whole and layered in salt in a large wooden box.

"The tenderloins were cold-packed in mason jars and kept in the cellar. Sausage was stuffed into cloth sacks, which Mama had sewn, and hung on the back porch. Sometimes some of it was fried, put into glass jars, then covered with lard. The head--including even the ears--was cooked, the meat pulled off and made into souse, a jellied meatloaf. The jelly had been obtained from the broth in which the head had been cooked.

"After Mama had cleaned and cleaned and cleaned the pigs feet, they were boiled. They were delicious with lots of raw onions on a cold winter's night. The livers were fried. They had to be eaten quickly and were often shared with neighbors. Hearts and tongues were boiled, sliced and eaten cold. Very little of our hog meat was wasted, but we didn't ever eat the blood. Some people did--they made blood

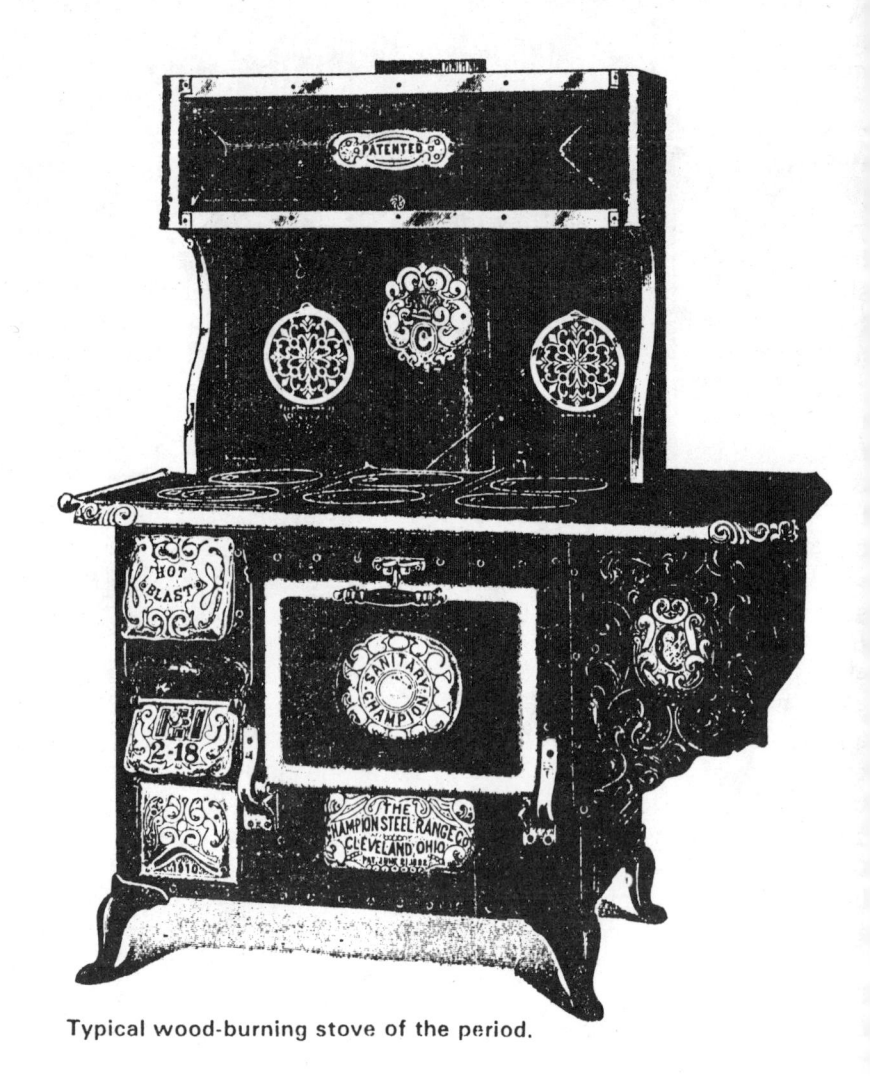

Typical wood-burning stove of the period.

pudding.

"We canned some fruits and vegetables obtained from our garden or purchased from neighbors who had raised a surplus. How I hated all the work involved! Stemming gooseberries and seeding cherries was never-ending drudgery. My arms were often sticky all the way to the elbows with cherry juice and I was thoroughly miserable.

"We had our own eggs, milk, cottage cheese, cream and butter. We raised potatoes and kept them in the cellar the whole year. Before a new crop was harvested the old ones were shriveled. I thought I'd go mad trying to peel enough of them to feed our hungry men.

"For years all of our bread was made at home. We had biscuits for breakfast and sometimes cornbread for other meals. Mama baked several loaves of what she called light bread twice each week. She had to start after supper to have it baked for dinner the following day. This was a job she hated. She said, "I hope some day I can buy our bread already baked." She was able eventually to do just that. She bought our bread already baked and already sliced! It's nice to think, even now, sixty years later, that Mama got her wish about buying bread to eat.

"We bought some foods in large quantities. I didn't know that flour was ever packed in smaller than fifty pound sacks, or sugar in less than one hundred pound bags. In the fall we bought apples by the wagon-load and stored them in the cellar. Every year we bought one hundred twenty pounds of sorghum and the same amount of honey.

"Fresh fruit, except for apples, was very rare in our house during the winter. We sometimes had oranges for special treats at Christmas. The first time I heard of grapefruit I was a teenager. I heard someone in a restaurant order half of one. I wondered why he wanted only half. I could only conclude that he wasn't very hungry. We had no fresh vegetables, either. Lettuce salads were unheard of in the winter. At Thanksgiving and Christmas we might have celery. Nor had we heard of taking vitamins. Still, we were as well as children today seem to be-- maybe more so.

"Mama was very proud of her reputation as a cook. Her snow-white layer cake was the first to disappear at family reunions. No one could make biscuits as light and fluffy as hers. Her boiled hams always drew raves, but her joy in this was diminished when someone pushed the fat to one side and left it on his plate. Horrors! One was supposed to eat every bite of it.

"Three times a day, six days a week, we all sat down together around the big table in the kitchen and had a hot meal. On Sundays Mama goofed off. We had our usual breakfast and an especially nice meal when we got home from church. But on Sunday night she set the leftovers out on the kitchen cabinet and we had to fend for ourselves. Papa would never cooperate with this arrangement. He sat down at the head of the table Sunday night the same as he did for every other meal of the week and ordered someone to bring his meal to him.

"We ate well and always had plenty available for company--expected or unexpected. In those days unexpected company was the common thing. There was no waiting to be invited--not even any phoning ahead. People--whole families of them--just dropped in. And we could feed them.

"My parents always seemed worried about money, and I felt that we were poor, but we were well fed. From the time Papa shook down the ashes in the stove to cook breakfast until we went to bed at night, our stomachs were full of good food."

## Keeping warm took real effort back in 1914

"Keeping warm was always a problem for my parents. I suppose it was a struggle that began in their childhood.

"The first report of it that I heard concerned Papa's battle with the cold when he was a bachelor. He had scraped up enough money to start farming and was living in a very crude house. When he read during the long winter evenings, he placed his chair on the dining table. It was so much warmer there than on the floor.

"I spent my childhood in a large, old frame house. It was during the years from 1914 until 1935. In the winter we had a stove in the living room and the kitchen stove was left up the year around. The rest of the house was unheated. The fire in the living room burned day and night, but that couldn't be managed in the cook stove. The kitchen often got so cold during the night that water froze. When the weather was severe, Mama moved her house plants into the living room.

"At such times our living room became very crowded. There were always several chairs and tables, as well as a piano and a secretary. The stove occupied quite a bit of space. During cold weather Mama often moved her sewing machine in and sometimes we had a quilting frame up for several months. It took a bit of squeezing to get it all arranged.

"We lived on a high hill where the cold winds came unimpeded from the North Pole. The idea of insulating had not been conceived. We had two large screened porches. Every fall Mama sewed canvas curtains, which we tacked over the screens. There were no storm doors, so canvas covered those screens, also. Mama folded newspapers and put them in the cracks around the unused doors. It took some ingenuity to keep warm.

"Keeping warm at night in our unheated bedrooms required some effort. We piled on layers of heavy woolen blankets. We slept on thick featherbeds. Mama heated her flat irons on the cook stove after supper. They were wrapped in old newspapers and put in our beds to warm our feet.

"I remember that at bedtime I took off my clothes and put on a long flannel nightgown behind the hot stove in the living room. The rest of the family very politely looked the other way. Papa carried me to bed so I wouldn't have to walk across the cold floor. He dropped me on the featherbed and I put my feet on the hot iron as he tucked the covers around me. How delicious it felt to be so snug in the frigid room.

"It was necessary to wear lots of warm clothing during the winter months. We had underwear with long sleeves and legs reaching to the ankles. It had to be folded to get our long stockings over it. I hated the lumpy look it gave my legs. Sometimes when I got to school, I went to the restroom and rolled the underwear up above my knees. That was after I got to

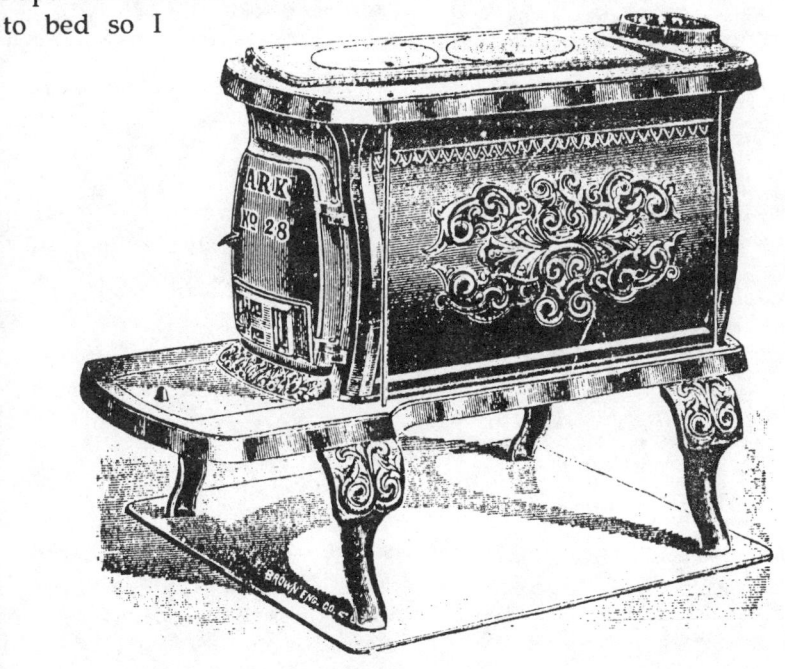

Houses in the early 1900's utilized stoves such as this one to heat individual rooms. The stove above doubled as a cooking surface.

high school. I went to grade school in the country and our restrooms were outside, and of course, unheated. It was no place to go to rearrange one's clothing.

---

**ONLY $37.50**

**PROTECTS YOU EVERY DAY**

If you drive a Ford runabout —1915-16 or 17 models, you need the new and classy, convertible

**Koupet Top**
TRADE MARK

The "Koupet Top" is made ONLY for the Ford runabout, but is similar in operation to the expensive touring Sedan tops. Never before has so good a top been offered at so low a price.

We have been making closed carriages since 1857. Our skill and experience is summed up in the Koupet Top. You will appreciate its style, quality, finish, and many other exclusive features.

The frame is of hard wood, covered with best quality 32-oz. rubberized duck. Side panels and doors are of glass and may be removed in a few moments. The "Koupet Top" is snug in blizzard weather and airy in midsummer.

The windshield is the newest double-acting, ventilating, automatic type. Both the doors and windshield are adjustable to any position by patented, self-locking devices. *They will not rattle.*

You can easily put the "Koupet Top" on your own car. No skilled labor required. It will outlast the car.

Write for circular or order at once if you are in a hurry. Weight 75 lbs. Shipping weight 110 lbs. Price F.O.B. Cars Belleville, $37.50. Money back if not satisfactory after 10 days use.

**Heinzelman Bros. Carriage Company**
116-25 Koupet Bldg.     Belleville, Ill.

---

"At that time cars were not closed in. (In fact, when closed cars came on the market, some people were afraid they would be too hot in the summertime.) Canvas curtains with isinglass peepholes snapped on open cars to keep out rain and cold. They were hard to keep in repair and sometimes we had to hold them in place. Cars had no heaters. Lap robes were standard equipment. There were devices similar to towel racks where we could hang them when they were not in use. Hot irons or bricks could be used to keep feet warm. When we were going to a family reunion, we put the roaster with the hot turkey on the floor of the back seat section. We children kept cozy with our feet on it.

"Our parents were obsessed with keeping warm and they had reason to be. It took considerable ingenuity, but they kept their family fairly comfortable in the big, old house on the windy hill during the coldest months of the year."

## Memories of an old storm cellar

"A cellar has long been an important part of a farm home. It is often called a storm cellar and manufactured ones are still installed in areas where severe storms frequently occur. It is a haven of safety if a tornado threatens. I have known families who retreated to the cellar during thunderstorms accompanied by high winds.

"The most common type of cellar, until very recently, was a cave. There was some sort of house over the entrance, usually with a door at both the top and bottom of a flight of steps. I remember as a child on our farm in the midwest, hearing lots of talk about whether or not one had a *good* cellar. Ours was considered *good* and to this day neighbors sometimes ask to store their potatoes in it.

Our cellar was used to keep foods cool in the summer and to keep them from freezing in the winter. We heard of at least one person who seemed to be confused about how this worked. Papa used to love to go over this story about him.

A neighbor bragged about what a good cellar he had, saying, "Our cellar is just wonderful. It is warm in the winter and cool in the summer."

"Papa said, "Don't you think that's just the way it seems because it's warmer than the outside air in winter and cooler than the outside air in summer?"

""No sir," insisted the neighbor. "It's warm down there in the winter and cool in the summer."

""Well," Papa agreed, "that certainly is remarkable. I don't suppose anything ever freezes in your cellar."

"The other man replied, "Well, it has to be awfully cold outside before things freeze down there."

"Papa always ended the story

there. I don't know whether his pupil learned the lesson or not. Papa was logical and his point was well made.

"When we were small, my brothers and I had fun playing around the cellar. There is a song that says, "Look down my rain barrel; slide down my cellar door." We slid down our cellar door. We rode our wagons and tricycles over the cellar hill. (Cellars were never completely underground, but were covered by a mound of dirt.) Now our grandchildren like to sit on the cellar roof.

"Years ago I played with a neighbor girl who took me down into their cellar. We took teaspoons with us and ate sour cream, straight out of a large stone jar.

"Once we found out that some men hired to do work on the furnace had helped themselves to a jar of peaches from our cellar. They had left an opened quart jar half full under the house.

"We always had potatoes, apples and five gallon cans of lard in our cellar. There were also shelves filled with literally hundreds of jars of fruits, vegetables, pickles and preserves we "put up" during the summer months. It was a dreaded chore to carry the milk and cream in stone jars up and down the cellar steps. That was usually my task before and after every meal. However, the cool milk tasted good, and cool cream was needed for making nice butter.

"Without refrigeration, a good cellar was almost a necessity. It is still about the only way to keep such things as large quantities of potatoes and apples for a long period of time. Some of our fondest memories center around our underground storage facility. We were fortunate to have a good cellar on our farm."

## I wanted to look like a town girl

Living in cities and large towns, we tend to forget what the "city vs country" competition was all about a few generations ago. With today's cookie-cutter fashions, and a discount store at nearly every rural intersection, the differences have grayed out. But they were there once, and Mildred S. Burns remembered them in this 1983 *Discover North* article.

"When I was a girl growing up on a Missouri farm in the 1920's, I often worried about the appearance of my clothes. I didn't want to look different from the town girls. Mama and I often disagreed about they way my clothes should fit. She wanted them longer than I thought they should be. I wanted them tighter than she thought they should be. We had a generation gap before anyone ever thought of the term.

"Mama made my dresses. She was a skillful seamstress, but I imagined my clothes looked homemade. Maybe they did, but she took pains to have them look professionally done. She had picoted or pleated ruffles and covered buttons and belts made by establishments that specialized in such things. Once, for a very important piano recital, she had the local milliner make a little flower out of dress material and pinned it at my waist. I remember a lavishly hand-embroidered collar and cuffs that she put on a black wool dress for me.

"We shopped in Kansas City and St. Joseph for coats, hats, shoes, gloves and purses. We spent a full day several times each year searching through the stores. Mama never bought anything until she had looked everywhere and compared prices. This involved lots of walking and, of course, we had to carry our packages. It took stamina to survive a day of shopping.

"All winter I had to wear cotton knit underwear with long sleeves and legs reaching to the ankles. I wore one pair a whole week. It got stretched and was very baggy before time for my Saturday bath and a clean suit.

"While Papa and the boys were outside working, I spread newspapers on the floor behind the stove in our big living room. That's where I had my weekly bath and put on clean underwear. It felt absolutely wonderful to be washed all over and have on fresh, snug-fitting underclothes.

"In the summertime I went barefoot at home. At all other times I wore long stockings held up with garters. In the winter the long underwear was folded over the stockings and looked unattractively lumpy. I didn't tell Mama that I rolled the legs of the longies up over my knees and put on smooth stockings as soon as I got to school.

"Mama usually made the garters that held up my stockings from elastic that she bought by the yard. In a pinch, I have used strips cut from old automobile inner tubes. Occasionally, I had a fancy pair that I received as a gift. They were made in pretty colors trimmed with lace, ruffles and bows.

"In the summertime I wore

bloomers. Mama made them. They were quite roomy. There was elastic around the waist and legs. They were made to reach just above the knee. Sometimes I pulled them up higher. How I hated for them to show, especially when they were made out of heavy black sateen. Sometimes they were made from old flour sacks with perhaps some printing still showing. There were times, too, when I was really dressed up with bloomers made out of the same material as my nice dresses. Then I was proud.

"I was the only girl in the family and Mama took pains with my clothes. Still, I often felt they didn't look just right and longed for ready-made dresses. That's the way it was when I grew up wishing I looked like the town girls."

## HERMITAGE BREAD PUDDING

2 cups sweet milk
1 heaping tbs. butter or oleo, melted
Pinch of salt
1/2 cup breadcrumbs  (toasted heels or slices of home made bread
1/3 cup raisins

1 cup white sugar
2 eggs, well beaten
1/2 teaspoon nutmeg

Scald milk and dissolve the sugar in it. When cool, add butter and eggs. Next beat in the salt and nutmeg. Grease a deep baking dish and line with the bread crumbs. Sprinkle raisins over crumbs. Pour custard mixture over crumbs and raisins. Set dish in pan of water and bake 10 minutes at 400º. Reduce heat to 300º and bake 30 minutes longer.

Frances Uttinger<br>Kansas City, Kan.

## FRIED SWEETINGS
*A Southern Recipe, from the 1920's*

6 sweet apples
Fat

6 tablespoons molasses

Wipe and core the apples and cut each in 3 or 4 rings. Heat a little fat in frying pan, lay the apples in cover and cook slowly, turning at the end of 15 minutes. When both sides are brown, pour one teaspoonful of molasses over each slice, cook five minutes longer and serve hot. Serves six.

Maxine Bowles<br>Knoxville, Ark.

## SOUP FROM ALSACE
*From a 1927 cook-book.*

6 cupfuls well seasoned stock
2 tablespoons flour

2 whole eggs
1 cupful cold water

Beat together the eggs and flour, adding the cold water *very slowly* until a smooth consistency is obtained. Pour the boiling stock over this mixture, *slowly* so as to avoid lumps, stirring constantly, and cook for five minutes in top of a double boiler, over hot water. This soup is most appetizing and makes a splendid dish to serve an invalid. Serves 10.

Maxine Bowles<br>Knoxville, Ark.

## APPLESAUCE CAKE

2 1/4 cups applesauce
   (unsweetened)
3/4 cups scant lard or butter
1/2 teaspoon nutmeg
3 tablespoons boiling water
3 cups flour
1 cup chopped nuts

1 1/2 cups sugar
1 1/2 teaspoon cinnamon
1/4 teasoon cloves
3 rounded teaspoons baking soda
1 1/2 cups raisins

Mix sugar, lard and applesauce well. Dissolve soda in boiling water, add to sugar mixture. Combine dry ingredients. Add with raisins and nuts to sugar mixture. Pour into 9x13 baking pan and bake at 350º. No time was given to me.

This recipe was given to me by a cousin of my husband in 1949. She told me it came from some of her family.

Gera Sawyer<br>Edgerton, Kan.

## YUM YUM BARS (DATES)

1 cup sugar
1/2 cup milk
2 tablespoons baking powder
1 cup chopped dates
powdered sugar

1 egg
1 cup flour
1/2 teaspoon salt
1 cup chopped nuts

Beat egg well, add sugar, mixing well. Combine flour and baking powder, and salt. Add milk and dry ingredients to sugar mixture. Fold in dates and nuts. Bake at 350° for 30 to 35 minuters.

While hot, cut into squares and roll in powdered sugar. Store in tight container so they will not dry out. Use a 9x9 baking pan.

This comes from a cookbook of the 1920's that my mother had. She prepared these for Christmas when I was a child. It was a treat at Christmas time.

Gera Sawyer
Edgerton, Kan.

## POOR MAN'S COOKIES

Put in 2-quart sauce pan:

1 cup water
1 1/2 cups seedless raisins

1 cup white sugar
1 stick oleo

Bring to a boil and let cool. Then sift together:

2 1/2 cups flour
1/2 teaspoon baking powder
1/2 teaspoon cloves (nutmeg can be used as substitute for cloves)
Dash of salt

1 teaspoon soda
1 teaspoon cinnamon

Add dry ingredients to cooled mixture, stir well; add 1 teaspoon of vanilla and 1 cup chopped nuts.

Spread batter on large size greased and floured cookie sheet. (Cookie sheet must have sides on it.) Bake at 350° for 12 to 15 minutes.

Ice while hot. Two cups of powdered sugar, milk and butter may be used for the icing.

Louise M. Bell
Utica, Ohio

## MOM'S CARROT CAKE

(Preheat oven to 350º)

1 1/2 cups sugar
1 1/2 cups shredded carrots
1/2 teaspoon salt
1 teaspoon cloves
1 teaspoon nutmeg

1 1/2 cups water
1 1/2 cups raisins
1/3 cup shortening
1 teaspoon cinnamon

Mix all together and boil ten minutes. Let cool. Then add:

2 1/2 cups flour
1 1/2 cups pecans, chopped fine.

1 teaspoon soda

Mix together. Pour in 13x9x2-inch pan. Bake about one hour at 350º or until an inserted toothpick comes out clean.
Use a cream cheese icing or serve with hot rum sauce. Serves 12. Great for gatherings or special occasions.
My mother brought this recipe from Kentucky in about 1914.

Arnold Cunningham
Harrisonville, Mo.

## KNEE PATCHES
*An old Swiss recipe.*

1 quart flour
1 cup sweet milk

3 eggs
Salt to taste

Add milk or flour as needed to make dough like pie dough. Take piece of dough about the size of a hulled walnut and roll out as thin as you can. Fry in deep fat to a golden brown. Immediately dip in a pan of sugar until well coated.
This recipe originated in Switzerland and was given to me by my late grandmother, Mrs. J.C. Schweizer. It is called "Knee Patches" because as the Swiss mother made them for her children, she would put a clean white cloth over her knee and pull and knead the dough over her knee until it had the correct consistency. We have had these yearly at the family gathering as long as I can remember. There is usually a stack of

them up to two feet high. They are eaten as snacks throughout the day, until there isn't a crumb left.

Kathleen Potter<br>
Savannah, Mo.

## AUNT ETHEL'S BANANA NUT BREAD

1 cup sugar
1 1/2 cup mashed bananas
1/4 cup shortening
1 1/2 cups flour

1/2 teaspoon soda
1/2 teaspoon salt
1/2 cup chopped nuts
3 beaten eggs

Preheat oven to 325º. Cream shortening and sugar, then add the eggs, bananas and nuts. Sift together the dry ingredients and add to the above. Pour into a greased loaf pan and bake for about 45 minutes or until done when tested with a toothpick.

Arnold Cunningham<br>
Harrisonville, Mo.

## BREAD PUDDING

(Preheat oven to 400º)

6 eggs, beaten lightly
4 cups (1 quart) milk, scalded
1 1/2 teaspoon vanilla
6 slices toasted bread (break into 1-inch pieces)
nutmeg

3/4 cup sugar
1/2 teaspoon salt
3/4 teaspoon lemon juice

1/3 cup raisins (optional)

Blend eggs, sugar and salt in large mixing bowl. Slowly stir in scalded milk. Add flavoring. Pour into a 13x9x2-inch baking dish over toast, broken into 1-inch peices. Push toast under with spoon and then sprinkle liberally with nutmeg.

Bake in 400º oven about 35 minutes or until inserted knife blade comes out clean. Serve warm or chilled. Drizzle honey over top or serve with hot rum sauce. Serves 12 to 14. Great for company.

Arnold Cunningham<br>
Harrisonville, Mo.

## MY GRANDMOTHER'S OATMEAL COOKIES

2 cups regular oats
1 cup sugar
1 cup raisins
1 scant teaspoon baking soda
  in 4 tablespoons milk
Few drops of vanilla

2 cups flour
1 cup shortening
2 eggs
1 heaping teaspoon cinnamon
Pinch of salt

Mix all together, drop by spoonfulls on ungreased cookie sheet. Bake 10 to 12 minutes at 375º.

Shirley J. Herbert<br>St. Joseph, Mo.

## BUBBLE BREAD

1 pkg. frozen dinner rolls (about 24 dough balls)
1 cup brown sugar
1 box butterscotch pudding   (NOT instant)
1/4 cup white sugar          1 teaspoon cinnamon
1/2 cup pecans               1/2 cup oleo (melted)

Grease and flour angel food cake pan. Place frozen rolls in pan.
Mix brown sugar and pudding mix and sprinkle over rolls. Mix white sugar and cinnamon and sprinkle over all the above. Sprinkle pecans over mixture and then pour oleo over all. Cover with plastic wrap and place on cabinet overnight.
Bake at 350º for 30 minutes or until done. Turn upside down on large plate and eat while warm.
This is a good quick breakfast.

Kathy Osborne<br>Sedalia, Mo.

## CHEAP FRUIT CAKE

An old fashioned cheap fruit cake, which was allowed to stand two or three days before it was baked, is the following:
Cream a cup of brown sugar, a cup of butter and a cup of milk, 4 eggs, 2 pounds of raisins, seeded and chopped fine, 1 grated nutmeg and a

tablespoonful of brandy; add 2 cups of sifted flour.

Beat the whole together and let it stand for two or three days in a cold place. When ready to bake it, sift 2 tablespoonfuls of cream tartar and 1 teaspoonful of soda together two or three times, with 2 cups of flour and stir it with the remainder of the cake. Bake in a slow oven for at least two hours.

"Another recipe from my grandmother's 1908 cookbook. Note the reference to "old fashioned" in 1908."

Faye Keyton
Independence, Mo.

## CREAMED ONIONS

2 pounds small white onions
1/4 cup shredded cheese
1 can cream of mushroom soup
1 tablespoon butter or margarine
1/4 cup cream or milk

Cook onions in lightly salted boiling water until almost done. Drain. Place onions in a 1 1/2 quart casserole, dot with butter. Stir soup in pan with cream or milk. Blend well. Pour over onions. Sprinkle cheese over top and bake at 350º for thirty minutes or until cheese is melted and slightly brown. Serves 6.

"I like fresh young onions, or spring onions, when they have a small bowl formed."

Mrs. O.L. Babcock
Platte City, Mo.

## GERMAN SHORTCAKE

A popular bread for breakfast is German shortcake. To 1 quart of flour add 1 tablespoonful of lard, 2 heaping teaspoonfuls of baking powder, 2 tablespoonfuls of sugar, a saltspoonful of salt. Mix all the ingredients with the flour, and add enough milk to make a soft dough. Roll into two sheets, put in the pie plates, allowing it to come up on the sides a little, spread with melted butter, sprinkle with granulated sugar and ground cinnamon.

"From a 1908 cookbook owned by my grandmother."

Faye Keyton
Independence, Mo.

## HOUSEHOLD REMEDIES, CIRCA 1908

**BUTTER FOR BRUISES.** While mothers' kisses are supposed to take the soreness out of all sorts of hurts and bruises, even this sovereign specific will not keep a bump from turning black and blue. If a bump is well buttered soon after it is made, the skin, it is said, will not change color. A woman who did not believe it, but tried it all the same, says her children have been saved many ugly marks by means of this simple and inexpensive remedy.

**SIMPLE REMEDY FOR CHIGGERS.** Put a few drops of carbolic acid into the water in which you bathe every morning, and you will have no more trouble with chiggers or seedticks.

**FOR JAUNDICE.** Take horseradish roots and grate them; put them in good cider vinegar and let the patient drink it.

**QUININE CURE FOR DRUNKENNESS.** Pulverize 1 pound of fresh quill-red Peruvian bark and soak it in one pint of diluted alcohol. Strain and evaporate down to one-half pint. For the first and second days give a teaspoonful every three hours. If too much is taken, headache will result, and in that case the doses whould be diminished. On the third day give one-half a teaspoonful; on the fourth reduce the dose to fifteen drops, then to ten and then to five. Seven days, it is said, will cure average cases, though some require a whole month.

**SOME HEALTH HINTS:** Sufferers with constipation might try to find relief as I do, by avoiding to eat what does "not agree," especially fats, new bread and pastry. Eat half a saucerful of boiled dried peaches, oatmeal and graham for breakfast; and if medicine is needed yet, take cascara tablets or cascara cordial. A cup of watery cocoa half an hour before breakfast is better than coffee. Eat buttermilk gravy for supper.

"The above are from a 1908 cookbook, "The Helping Hand," owned by my grandmother."

Faye Keyton
Independence, Mo.

## PARSNIP STEW

1/2 pound salt pork, cubed
2 onions, medium
2 to 3 cups hot water
4 parsnips, sliced

3 to 4 potatoes, sliced
1/2 cup (or more) rich millk
Parsley

Fry salt pork slowly, do not brown. Add onions, cook until yellow. Add water, stirring up all the goodness in pan. Add parsnips. Simmer covered 15 to 20 minutes, until parsnips are nearly done. Add potatoes and finish cooking. Salt and pepper to taste. Thicken slightly, adding milk. Sprinkle with parsley.

Serve hot, with hot biscuits or crunchy corn bread. Green salad and fruit dessert go well with this turn-of-the-century recipe.

Sandy Bare
Olathe, Kan.

## FAT ONIONS

6 medium Bermuda onions, cooked
4 medium potatoes, cooked & mashed

1 small cauliflower, cooked
Cream sauce

Scoop out onions, stuff with mashed, seasoned potatoes, put in casserole. Break up cauliflower, put it and scooped out onion around stuffed onions. Pour cream sauce over all. Put in oven until piping hot. Used by our family at the turn of the century..

Sandy Bare
Olathe, Kan.

## CHRISTMAS OYSTER SALAD

(1) 10-ounce can of oysters
1 cup chopped sour pickles
1 cup chopped celery
1/2 cup cracker crumbs

4 hard-boiled eggs, chopped fine
1 small jar chopped pimiento
Salt, pepper, celery seed to taste

Mix with mayonnaise. Another family recipe from around 1900.

Sandy Bare
Olathe, Kan.

## NEVER FAIL PIE CRUST

2 1/2 to 3 cups flour
2 teaspoons sugar
1 egg yolk
1/2 cup water (scant)

1/4 teaspoon salt
1/4 teaspoon baking powder
3/4 cup lard

Sift together all dry ingredients. Cut in lard until mixture is like meal. Add egg yolk and water. Turn onto board and knead lightly (working in extra flour as needed.) Form dough into a ball. Use what is needed.

Store remaining dough in refrigerator. Will keep for at least two weeks. Makes two shells or bottom and top crusts.

Mrs. O.L. Babcock
Platte City, Mo.

## CORN CHOWDER NO. 2

1 quart raw sweet corn
1 pint diced potatoes
1 teaspoon salt
1 saltspoon of pepper
1/4 cup butter

2 tablespoons flour
1 pint milk
2 hard boiled eggs
1 pint croutons

Cut each row of kernels, and scrape the raw corn from the cob. Boil the cobs twenty minutes in water to cover. Pare and cut the potatoes into small dice. Pour boiling water over them, drain and let them stand while the corncobs are boiling. Remove the cobs, add the potatoes, salt and pepper. When the potatoes are nearly done, add the corn and milk and cook five minutes. Cook the flour in the hot butter, add one cup of the corn liquor, and when thick stir it into the chowder. Add the eggs, whites chopped fine, and yolks rubbed through a strainer. Serve with croutons.

Delores Harrington
Tarkio, Mo.

## CAKE WITH NUT FILLING

Cream 1 cup of butter with 2 cups of sugar as lightly as possible. Add 1/2 teaspoonful of salt and 1 cup of finely chopped nuts. Add the grated rind and the juice of 1 lemon, and add gradually 4 cups of sifted flour, and the stiffly beaten whites of 4 eggs.

Next add 1 cup of milk, in which is dissolved 1 teaspoonful of soda. Sift in 1 teaspoonful cream tartar; beat all lightly and bake in jelly cake tins.

For the filling: To the white of 1 egg add a little salt and enough powdered sugar to make a soft icing. Set aside enough to cover the outside, and to the rest add a cup of finely chopped nuts and the rind and juice of 1 lemon. Spread this thickly between the layers. On the outside put the reserved icing, to which add a little lemon juice, and ornament the top with the perfect halves of the nuts. This is a very delicious cake.

"From a 1908 cookbook owned by my grandmother."

Faye Keyton
Independence, Mo.

## TEA PUNCH

4 quarts boiling water
1 quart cold water
2 cups fresh lemon juice
4 cups powdered sugar
1/2 cup tea
10 cups pineapple juice
Ice

Let boiling water stand on tea for 6 minutes. Then strain, add sugar, stir well till sugar is dissolved. Add cold water, pineapple juice, lemon juice, and chill. Serve in a large bowl over a block of ice. Makes 8 quarts before adding ice. Corn syrup may replace half the sugar. This recipe has been in the family since the early 1900's.

For a special treat, serve in a frosted rim glass.

## To Frost Glasses:

Put some lemon juice in a saucer about 1/4 inch deep. Sift some powdered sugar onto a plate, about 1/4 inch deep. Stand each glass in lemon juice (inverted). Then stand glass (inverted) in powdered sugar for a minute. Lift carefully out of sugar so as not to jar the sugar coating which has formed on the rim. Stand in ice box (refrigerator) until set. Then carefully fill with your drink.

Bernice McCumons
Marlette, Mich.

## BUTTER-LESS, EGG-LESS, MILK-LESS CAKE

1 cup brown sugar, firmly packed
1/3 cup vegetable shortening or lard
1 teaspoon baking powder
2 teaspoons water
2 teaspoons cinnamon
1 teaspoon salt
1 1/2 cup water
2/3 cup raisins
1 teaspoon soda
2 cups all-purpose flour
1/2 teaspoon powdered cloves
1/2 teaspoon nutmeg

Boil brown sugar, 1 1/4 cup water, shortening, raisins and spices together for 3 minutes. Cool. Add salt and baking soda which has been dissolved in 2 teaspoons water. Gradually add flour and baking powder, which have been sifted together, beating smooth after each addition. Bake in a greased and floured 8x8x2-inch pan in a moderate oven of 325º for about 50 minutes, or until done. Needs no frosting.

*Bernice McCumons*
*Marlette, Mich.*

## AUNT REBA'S SUGAR COOKIES

1 stick oleo (melted)
1 1/4 cup sugar
1 teaspoon vanilla
1/4 teaspoon salt
1 teaspoon cream of tartar
1/2 cup shortening
1 egg
2 1/2 cups flour
2/3 teaspoon soda

Beat oleo and shortening. Add sugar and egg, beat. Add remaining ingredients. Roll in balls and lightly flatten with sugared glass bottom. Bake at 350º until golden.

*Kathy Osborne*
*Sedalia, Mo.*

## GRANDMA'S CREAM PUFFS

Boil together 1/2 cup of butter and 1 cup water with a pinch of salt. While boiling, add 1 cup flour. Stir until smooth.
When still warm, add 4 eggs one at a time, not beaten. Stir smooth--it takes quite a bit of stirring.
Drop by spoonfuls on a greased baking sheet 2 inches apart. Bake in a

hot oven at 450º for first 15 minutes, then reduce heat to 325º for 20 minutes, or a little more. Do not open oven during first 20 minutes. Take one out and if it doesn't fall, it is done.

The high heat makes it puff and lower heat dries them out, so the puff is not soggy. You make a hole in the puffs to put in the filling. Use a paring knife.

### Filling for Cream Puffs

2 cups milk
3 eggs
4 tablespoons cornstarch

1 cup sugar
2 tablespoons butter
Vanilla

Heat milk and 1/2 cup of sugar and butter in a double boiler to the boiling point. Mix 1/2 cup sugar, cornstarch and beaten eggs, add to boiling liquid.

Cook until smooth and thick, stirring often. Add vanilla, cool. Don't beat eggs too much.

Alta Mae Morris
Liberty, Mo.

## GRANDMA'S DELICIOUS SUGAR COOKIES

Measure and sift together:

5 cups of flour
1 teaspoon salt

1 teaspoon soda

Beat together until light and fluffy:

1 cup oleomargarine
1 cup sour cream
2 eggs

2 cups sugar
1 teaspoon vanilla

Add dry ingredients, stirring well. Chill dough for several hours in a covered bowl in refrigerator. Roll out a small amount at a time on lightly floured board.

Cut with cookie cutter. Place on ungreased cookie sheet. Bake in hot oven (375º) for about 10 minutes.

Decorate with colored sugar and chocolate sprinkles. Makes 6 dozen.

Alta Mae Morris
Liberty, Mo.

## DROP DOUGHNUTS

1 cup sugar
1 tablespoon butter
3 cups sifted flour
1/2 teaspoon salt

2 eggs (well beaten)
1 cup milk
2 teaspoons baking powder
1 tablespoon vanilla

Mix all together and drop by small teaspoonful into hot fat or oil  Fry until brown on both sides. Roll in cinnamon sugar or powdered sugar.

Wilma Hays
St. Joseph, Mo.

## DIVINITY FUDGE

2 1/2 cups of sugar
1/2 cup water
1 teaspoon vanilla

1/2 cup corn syrup
2 egg whites, beaten stiff
1 cup chopped nuts

Cook sugar, corn syrup and water until it will spin a thread.  Then slowly pour half of mixture into the egg whites. Cook the remaining half until it will harden when tested in cold water. (A small amount will break when it is tested in water.)

Then, pur into the first mixture. Beat until creamy; add nuts to taste, and then vanilla. Pour into buttered dish.

"This dates back to the late 20's, but is still as good today."

Florney Gibbons
Kansas City, Mo.

## SKILLET MACARONI

Use 1 1/2 cups of elbow macaroni. (Shells were a treat.) Brown in skillet in oleo or bacon fat. Will need to shake or stir to prevent burning.

At desired shade of browning, pour water in skillet. Add 1/2 pound ground meat, chopped pepper and onion. Cook until macaroni is done. Add seasonings and serve.

If aunts, uncles and cousins show up, just add more (browned) macaroni and put in a soup pot. Then add a quart of tomatoes to the other meat, peppers and onions when half-done. This makes a full meal with just a lettuce salad.

June Medina
Baldwin City, Kan.

## SOUR CREAM RAISIN PIE

1 tablespoon flour
1/2 teaspoon cinnamon
2 egg yolks
1 1/2 cups raisins

1/2 teaspoon nutmeg
1 cup sugar
1/2 cups smooth sour cream
1 unbaked pie shell

Mix the flour, spices and sugar with the slightly beaten eggs. Add the sour cream and raisins and pour into pie shell. Bake at 350º for about 40 minutes. Beat the two egg whites until stiff peaks form. Add 1/4 teaspoon cream of tartar and two teaspoons sugar. Cover pie and seal edges. Return to oven until browned. This recipe came from one of my aunts, who used it in the early 1900's.

Shirley Stafford
Edgerton, Mo.

## SWISS PUDDING
### By Crystal Adkins

4 cups sponge (see below)
Flour to make stiff dough
1/2 cup water
Cinnamon to taste

1 1/4 cups sugar
4 cups raisins and currants
Salt to taste
Cut-up dates, if desired

### SAUCE:

Grape juice or wine thickened with cornstarch and sweetened to taste.

### SPONGE:

1 cake yeast dissolved in
    1/4 cup warm waterr
2 tablespoons shortening

2 tablespoons shortening
4 teaspoons salt
4 cups liquid

Mix with enough flour to make sponge and let set overnight.
Mix pudding to desired stiffness. Place in muslin bag sewed the shape of a large ham. Place in large kettle, in boiling water, and boil two hours.

It can also be steamed.

This recipe has been a Christmas dinner dessert as far back as I can remember. Before electricity and gas invaded the kitchens, I can remember my mother putting the bag of pudding in a copper boiler and boiling it on top of the range. The pudding is the highlight of the family dinner.

Kathleen Potter
Savannah

## 'HAND-ME-DOWN' CHOCOLATE CAKE

3/4 cup butter (or margarine)
2 eggs
2 cups all-purpose flour
1 1/4 teaspoons baking soda
1 1/3 cups water

1 3/4 cups sugar
1 teaspoon vanilla
3/4 cups Hershey's Cocoa
1/2 teaspoons salt

Cream butter and sugar until light and fluffy. Add eggs and vanilla, beat one minute at medium speed. Combine flour, cocoa, baking soda and salt; add, alternating with water, to creamed mixture.

Pour batter into two greased and floured 8" baking pans. Bake at 350º for 35 to 40 minutes. Cool, frost with chocolate frosting and decorate.

"This dates back to the late 1920's when our mothers had the old oil cloth for our kitchen table...and our eyes got as big as the cake in anticipation of just a small piece of the cake."

Florney Gibbons
Kansas City, Mo.

## EGG BUTTER

3 cups dark syrup
4 eggs

2 sticks (1 cup) margarine or butter

Beat the eggs as light as you can get them. This is very important.

In the meantime, bring the syrup and margarine to a rolling boil, using a large kettle. Pour the egg mixture into the syrup, using a thin stream and stirring constantly, until every vestige of foam disappears. Pour into containers. This should be the consistency of peanut butter and can be kept a long time in a refrigerator. It can be used as a spread for bread, a topping for ice fream, or a filling for cakes.

"My mother-in-law made this often and always in the spring, when eggs and butter were plentiful on the farm."

Edith Snider
Kansas City, Mo.

## OLD FASHIONED PORK SAUSAGE

My parents, the late Harry and Nell Martin of DeKalb, Mo., used this sausage recipe when they killed hogs down on the farm. Nowadays you can have your butcher grind pork for you to make your own sausage. For 10 pounds ground pork:

2 1/2 tablespoons salt
1 tablespoon red pepper
1 tablespoon black pepper

Mix well. Let set awhile to season good. They cured the fresh hams, shoulders and bacon with this sugar cure:

2 cups brown sugar
1 tablespoon red pepper
2 cups canning salt
2 tablespoons black pepper

Mix well, rub on the fresh pork. Keep in cool place to cure. I use this on fresh sliced pork shoulder. It keeps good for quite awhile in the refrigerator.

Jessie Black
St. Joseph, Mo.

## PUMP 'CAN' BREAD

3 cups flour
1 teaspoon salt
1/2 teaspoon nutmeg
1/2 cup chopped pecans
1 tablespoon baking powder
1 teaspoon cinnamon
1/4 teaspoon ground cloves

Mix all that well. Now beat 2 eggs, and add 1 cup milk, 1 cup pumpkin, 3/4 cup light brown sugar, 1/4 cup oil, 1 teaspoon baking soda. Add this to the first mixture, stirring well. Pour into two well greased 1-pound coffee cans.

Preheat oven to 350º and cook 1 hour and 15 minutes. Let cool about 20 minutes before removing from cans.

Janice Miller<br>Oraville, Calif.

## EVERY DAY COOKIES

4 eggs
2 cups sugar
1 teaspoon soda
2 teaspoons baking powder

1/3 cup sour milk
1/2 teaspoon nutmeg
1 cup shortening (lard at that time)

When mixing ingredients, add enough flour (about three cups) to make a stiff dough. Roll out and cut into desired shapes. Bake in moderately hot oven (375º).

"When I clipped this recipe some time back it was said to have been the recipe of Mrs. Cessna, the mother of Clyde Cessna, the airplane pioneer. The recipe was used regularly in the early 1900's.

I sometimes just rolled dough log fashion and sliced off and baked when in a hurry. My two boys loved them. I made these often in the 60's."

Lulua Fillmore<br>Independence, Mo.

## MUSH BISCUITS

1/3 cup corn meal
1 teaspoon salt
2 cups milk
1 cake yeast
4 cups flour

1/2 cup sugar
1/2 cup shortening
2 eggs (whites beaten)
1/4 cup lukewarm water

Mix corn meal, sugar, salt and shortening in a double boiler. Cook until thick, stirring frequently. Cool mixture to lukewarm, then add egg whites and yeast (which has been softened in warm water).

Beat thoroughly and let rise in warm place one hour. Then add four cups of flour, one at a time. Knead on floured board, adding no more than one more cup of flour during the kneading.

Place in greased bowl. Let rise to double size. Make into rolls and bake at 375º for fifteen minutes. This dough can be kept for several days in the refrigerator and baked when you need it.

This is an old recipe of my mother, Mrs. N.L. Farmer. It is good, and different from the other rolls.

Mrs. O.L. Babcock<br>Platte City, Mo.

## AUNT KATIE'S QUICK INDIAN PUDDING

3 cups milk
1 tablespoon yellow corn meal
1/2 cup molasses

1/3 cup quick-cooking tapioca
3/4 teaspoon ginger
1 tablespoon butter

Place 2 cups milk in top of double boiler. Add tapioca gradually. Cook 15 minutes, stirring frequently. Add corn meal. Cook 15 minutes more. Remove from heat. Add ginger, molasses and butter. Mix well. Pour into buttered baking dish. Pour remaining milk over tapioca mixture. Do not stir. Bake in 325º oven for 3 hours. Serve with a pitcher of heavy cream.

Emmy Fleming<br>Leawood, Kan.

## RUM TUM TIDDY

1 can condensed tomato soup
1/2 lb. sharp Cheddar cheese
1/4 teaspoon dry mustard

1 egg slightly beaten
Buttered toast slices

Pour soup in double boiler. Add grated cheese. Cook over hot water, stirring constantly until blended. Remove from heat. Add mustard and slightly beaten egg. Mix well and serve over buttered toast points.

Emmy Fleming<br>Leawood, Kan.

## POTATO CAKE

2/3 cup butter
Yolks of 4 eggs
2 squares chocolate
2 cups flour
1 teaspoon cinnamon
1/2 teaspoon ground cloves
Whites of 4 eggs

2 cups sugar
1 cup of hot mashed potato
1/2 cup milk
3 1/2 teaspoons baking powder
1 teaspoon nutmeg
1 cup chopped walnut meats

Cream together butter and 1 cup of sugar, beat to a froth yolks of eggs with remainder of sugar, then blend both mixtures thoroughly together. Add potatoes, chocolate melted over hot water, and alternately milk with flour, which has been sifted with baking powder and spices. Last, add whites of eggs whipped to a stiff froth and walnut meats. Bake in layers or a loaf cake as desired and cover with a chocolate or a white frosting.

"This is an old recipe from *Mrs. Curtis's Cookbook,* 1908, that my grandmother gave me in 1942 as a keepsake. There were no baking times or temperatures. I bake the cake at 350º for 30 to 35 minutes."

Iona Geib
Mound City, Mo.

## QUICK CAKE

1/3 cup soft butter
2 eggs
1 3/4 cups flour
1/2 teaspoons cinnamon
1/2 pound dates stoned and cut in pieces

1 1/2 cup brown sugar
1/2 cup milk
3 teaspoons baking powder
1/2 teaspoons grated nutmeg

Put ingredients in a bowl and beat all together for three minutes, using a wooden cake spoon. Bake in a buttered and floured cake pan 35 to 40 minutes in a 350º oven. If directions are followed this makes a most satisfactory cake. But if the ingredients are added separately, it will not prove a success.

"This recipe came from *Mrs. Curtis's Cookbook,* 1908, that my grandmother gave me in 1942. The original recipe contained no cooking times or temperatures, and referred to "cupfuls, and teaspoonfuls.""

Iona Geib
Mound City, Mo.

## FROZEN FRUIT SALAD

1 1/2 cup sour cream
1/8 teaspoon salt
1/2 cup pecans (broken)
1 can No. 2 drained pineapple

3/4 cup sugar
2 tablespoons juice
1/4 cup cherries
2 large bananas, diced

Stir well and place in 13x9 pan and place in freezer overnight. Let thaw 5 to 10 minutes before serving.

Rosamary Lindsey
Orrick, Mo.

## JELLIED VEAL LOAF

3 lbs. veal
1 tablespoon butter
1/2 cup cold water
Salt and pepper to taste

1 large onion, chopped
1 tablespoon gelatin
2 stocks celery, chopped

Cut veal in pieces, add onion, celery and butter. Season to taste. Cover with cold water and cook slowly until meat is tender and liquid is reduced to about two cups. Soak gelatin in cold water (1/2 cup) about 5 minutes. Grind the veal and strain the stock. Dissolve the gelatin in hot stock and add the ground veal. Mix well and pour into glass loaf pans to jell.

This was a favorite recipe of my great-grandma's.

Emmy Fleming
Leawood, Kan.

## AUNT OPAL'S BROWN SUGAR PIE

1 cup brown sugar
1/2 cup cream

1/4 cup butter
2 eggs

Beat yolks lightly. Add sugar, cream and vanilla, and melted butter. Beat whites of eggs to stiff froth, fold in. Pour mixture into pastry shell and bake well. This is another favorite family recipe from around 1900.

Beverly Rizzo
Kansas City

## BAKED MACARONI AND TOMATOES

2 cupfuls cooked macaronii  
1/4 teaspoonful salt  
Buttered crumbs  

1 can or jar of tomatoes  
2 tablespoonfuls melted fat  

Add fat and salt to the tomatoes, arrange in layers with macaroni in a greased baking-dish. Cover with buttered crumbs. Bake at 500º for 12 minutes. If very cold, bake at 400º until heated through and crumbs are brown.

I remember my Mom fixing this real often. We seldom had meat or cheese. I don't see it served anymore. It comes from a 1927 cookbook and was used by many during Depression days.

Maxine Bowles  
Knoxville, Ark.

## JERUSALEM PUDDING

(1) 3-oz. package lemon gelatin, regular  
1/3 cup raw rice, cooked in 1 cup water for 30 minutes  
3/4 cup crushed pineapple            1 cup of cream, whipped  
1 1/2 cups hot water  

Dissolve gelatin in hot water. Put in refrigerator until thick. Fold in rice, pineapple and whipped cream. Let firm up again in refrigerator and serve.

I have substituted whipped topping of late. Do not care for instant rice, though.

This recipe comes from my mom's 1920 cookbook. It is the dessert we had at Christmastime while I was growing up.

Gera Sawyer  
Edgerton, Kan.

## RICE AND CABBAGE SOUP
*From a 1927 Cook-Book*

3 cups shredded, half cooked cabbage  
2 quarts stock  
Parmesan cheese  

Fat  
1/2 cup rice, uncooked  

Saute the partially shredded cabbage in hot fat until golden brown.

Wash the rice and boil until tender in the stock, which should be very well seasoned. Add the cabbage, let them boil together for a few minutes. Pass freshly grated Parmesan cheese when soup is served. Seves 6.

Maxine Bowles<br>Knoxville, Ark.

## CRANBERRY JELLY
*Another 1927 recipe*

4 cups cranberries
1 cup boiling water

2 cups sugar

Pick over and wash cranberries. Place in a stewpan with boiling water. Boil until all cranberries have burst open, about 10 minutes.

Pour into a sieve and mash through as much of the pulp as possible. Add two cupfuls sugar, return to fire and bring slowly to a boil, stirring constantly so that all sugar is dissolved. Pour at once into wet molds or heated jelly glasses. Serves six to eight.

Maxine Bowles<br>Knoxville, Ark.

## FAVORITE CREAM CAKE

Break 2 eggs into a cup and fill up cup with cream. If sour cream is added, use 1/2 teaspoon soda. If sweet, use 1/2 teaspoon baking powder, 1 cup sugar and 1 1/2 cups flour. Beat all together in moderate oven till done.

Beverly Rizzo<br>Kansas City, Mo.

## OATMEAL BREAD

1 cake yeast
1/2 cup lukewarm water

1/4 teaspoon salt
1 tablespoon sugar

Crumble yeast into warm water. Add sugar and salt. Let stand until yeast is dissolved. In large mixing bowl put:

1 cup regular long-cooking oatmeal
1 teaspoon salt
Some chopped nuts, if desired

3/4 cup brown sugar
1 cup raisins
2 tablespoons shortening

Over this pour 2 cups boiling water. Cook over medium heat until lukewarm. Add dissolved yeast mixture and 4 1/2 cups flour, or enough to make a soft dough. Let rise until doubled. Place in greased loaf pans. Let rise again. Bake at 400º for 15 minutes. Lower heat to 350º. Bake 30 minutes longer. Brush tops with glaze.

## GLAZE

2 tablespoons sugar

2 teaspoons water. Mix to blend.

Dorothy Krueger
Ripon, Wis.

## PEANUT BUTTER COOKIES

1 cup brown sugar
1 cup shortening
1 cup peanut butter
1 teaspoon vanilla
1 teaspoon soda

1 cup white sugar
2 eggs, beaten
3 cups sifted flour
1/2 teaspoon salt

Cream brown sugar, white sugar and shortening until fluffy. Add peanut butter, then eggs, mixing well after each. Add vanilla, then sifted dry ingredients. Mix well. Make into balls the size of a walnut.

Put on greased baking sheet. Flatten with fork to make criss-cross design. Bake at 350º about 10 minutes, or until lightly browned. Cool on rack.

Dorothy Krueger
Ripon, Wis.

## GERMAN RHUBARB PIE

3 1/2 cups chopped rhubard
3 beaten eggs
1/4 cup light cream

1 1/2 cups sugar
1/8 teaspoon salt
1 tablespoon flour

Put in pie crust. Bake at 375º until set--about one hour.

Alta Mae Morris<br>
Liberty, Mo.

## MINCE MEAT

Years ago my parents always butchered their own hogs. Neighbors would help each other as several hogs were butchered at one time and they always made their own mince meat, which made a most delicious pie.

4 pounds cooked pork cored)
2 tablespoons ginger
2 tablespoons cloves
6 tablespoons cinnamon
2 oranges
3 teaspoons salt

15 or 20 apples (peeled and
2 pounds brown sugar
1 pint molasses
3 pints boiled cider
5 teaspoons nutmeg

5 cups of mixed fruit, such as white and dark raisins, currents, dried peaches, apricots, chopped prunes.

Put this mixture into a large container and cook slowly together until fruits are done and flavor is mixed thoroughly.

Mrs. Eleanor Frakes<br>
St. Joseph, Mo.

## COTTAGE PUDDING

This recipe has an unusual name, but the origin is unknown. It was one of my mother's favorite recipes, and was a favorite when I was a child 60 years ago. It's especially good served warm, with the lemon sauce.

1/3 cup butter
2/3 cup sugar

1 3/4 cups flour
1/4 teaspoon soda

1 egg
1 cup sour milk
2 teaspoons baking powder
1/4 teaspoon salt
1 teaspoon vanilla

Cream butter, egg and sugar together. Sift flour, baking powder, soda and salt together, add to cream mixture alternately with the sour milk. Add vanilla and beat hard. Bake in greased pan at 350º for 25 to 30 minutes.

## Lemon Sauce

1 cup sugar
2 tablespoons butter
2 tablespoons flour
1 cup boiling water
2 tablespoons lemon juice or extract
Dash of salt

Mix sugar, flour and salt together. Add boiling water and lemon juice. Add the butter. Cook in double boiler pan until done. Serve hot over the cottage pudding.

Mrs. Eleanor Frakes
St. Joseph, Mo.

## FLAPPER PUDDING

This recipe was taken from a magazine in 1929. In those days it was hailed as the "cat's pajamas."

1 cup fine vanilla wafer crumbs
2 cups sifted confectioners sugar
2 stiff, not beaten, egg whites
(1) 9-oz. crushed pineapple, drained
3/4 cups butter or margarine
2 egg yolks
Have eggs at room temperature
1/2 cup chopped black walnuts
   (or choice)

Spread half of vanilla wafer crumbs on bottom of 10x6 1/2 baking dish. Cream butter gradually add sugar, beating until light and fluffy. Add egg yolks, one at a time, beating well after each addition.
Beat one minute more, fold in egg whites. Mixture may look curdled. Beat at medium speed for a few seconds, or until smooth, fold in pineapple and nuts. Carefully spread over crumbs, top with remaining crumbs. Chill until firm, 5 hours or overnight. May garnish with cherries.

Maxine Bowles
Knoxville, Ark.

## SWEET SOUR PORK

My mother made this for many years, and served it in a restaurant she owned in Logan, Iowa. Take 3 pounds of pork steak, cut in one-inch pieces and saute. Make sauce of:

1 1/2 cup brown sugar
1 cup vinegar
1/2 cup soy sauce

1/2 teaspoon ground ginger
2 cups water

Simmer in sauce. Just before you serve this, thicken it with 3 tablespoons of corn starch. Then add one large can of chunk pineapple, slices of green pepper, and fresh tomatoes in small chunks. Let these heat, not cook. Serve on rice.

Kathalyn Iversen
Mondamin, Iowa

## CHIFFON CAKE FOR BLARNEY STONES

Sift together in bowl:

2 cup flour
3 teaspoons baking powder

1 1/2 cup sugar
1 teaspoon salt

Put in bowl. Make a hole in middle of ingredients and add:

1/2 cup vegetable oil
3/4 cup vanilla

5 egg yolks
Cold water

Mix well. Fold the above mixture into the egg whites, adding 1/2 teaspoon of cream of tartar. Bake in ungreased pans (you'll need a 10x15 and one 8x8) at 375º until raised and lightly brown--about 45 minutes--springs back to touch.
You need to cut cake in bars like a "nickel candy bar." Cool cake on inverted glasses.

### Frosting for Blarney Stones

3/4 pound butter softened & creamed
3/4 cup cream

2 pounds powdered sugar
Vanilla

Beat until creamy--you may need more cream. Grind 2 pounds of Spanish peanuts. Pour from one pan to another to get rid of skins on peanuts. Frost each piece of cake on all sides and roll in peanuts.

Kathalyn Iversen<br>Mondamin, Iowa

## CHEAP AND EASY HOT SAUCE

1/4 cup butter             3/4 cup sugar
2 tablespoons flour

Cream the above ingredients, then add 1 1/2 cup boiling water. Stir well until thickened, and add vanilla. Put vanilla wafers in a sauce dish, and pour this sauce over them. This was a childhood favorite.

Kathalyn Iversen<br>Mondamin, Iowa

## MOM'S APPLESAUCE CAKE

2 cups sugar
2/3 cup butter
2 1/2 teaspoon baking soda
1 cup seeded raisins
1 teaspoon nutmeg
1 teaspoon vanilla
1 cup nutmeats

2 cups applesauce
1 egg, beaten
1/2 cup hot water
1 teaspoon cinnamon
1 teaspoon cloves
3 cups flour

Cream butter, sugar and egg. Add applesauce, dry ingredients, nuts and raisins. Another three-generation recipe that's still a favorite.

Beverly Rizzo<br>Kansas City, Mo.

## SANDWICH SPREAD

Grind enough green tomatoes to make a pint (without juice). Grind 2 green peppers and 2 red peppers. Mix tomatoes and peppers and sprinkle with salt. Let sit a few minutes and drain.

Put in pan and add 1/2 cup of water, cook until tender. Add 1/2 dozen ground sweet pickles. Keep hot until following dressing is prepared:

1 cup sugar
2 tablespoons prepared mustard
1 cup sour cream

2 tablespoons flour
1/2 cup vinegar
3 eggs, well beaten

Cook well. Then mix with tomatoes and seal while hot.

Mrs. Charles V. Brokenicky<br>Smithville, Mo.

## BLACK WALNUT PIE

1 cup brown sugar
1/2 cup black walnuts
3 tablespoons corn starch

2 egg yolks
1 1/2 cups sweet milk
1/2 teaspoon vanilla

Heat milk in double boiler. Mix cornstarch and sugar, and stir into milk. Add beaten egg yolks. Remove from heat and add black walnuts and vanilla. Beat egg whites, adding 1 teaspoon baking powder and 4 tablespoons of sugar. Brown meringue in oven.

Mrs. Charles V. Brokenicky<br>Smithville, Mo.

## BUTTERSCOTCH PIE

1 cup milk
1 cup brown sugar
1 tablespoon cornstarch
Butter (size of egg)

2 cups water
3 tablespoons flour
3 egg yolks
1 teaspoon vanilla

Boil water and milk together in double boiler. Mix all other dry ingredients together and stir in milk and water. Add the egg yolks and butter and cook until thick. Top with beaten egg white meringue.

This recipe came from a 1922 DeKalb Commerce Club cookbook belonging to my mother.

Dorothy Beisinger<br>DeKalb, Mo.

## BLACK CHARLEY CAKE

1 1/2 cups sugar
2 egg yolks
1/2 cup sweet milk
1 teaspoon soda
   in 1 tablespoon hot water

1/2 cup lard
2 tablespoons cocoa
   in 1/2 cup hot water
2 cups flour

Beat two egg whites until stiff and gently fold in last. Cream sugar and lard, and add egg yolks. Alternate liquids and dry ingredients, and fold in egg whites. Bake in moderate oven. Another favorite family recipe for three generations.

*Beverly Rizzo*
*Kansas City, Mo.*

## SNOW ICE CREAM

2 cups milk or cream
1 1/2 cup sugar

2 eggs

Mix ingredients after beating eggs. Add vanilla and snow.

*Beverly Rizzo*
*Kansas City, Mo.*

## MINCEMEAT

5 cups ground meat,
5 cups raisins
1 quart vinegar
1 teaspoon salt
1/2 teaspoon nutmeg
1/4 teaspoon cloves

10 cups chopped apples
5 cups sugar
1 cup chopped suet
2 teaspoons cinnamon
1/2 teaspoon allspice

Brown meat, add rest of ingredients and cook until apples are done or it has thickened. Can or freeze what is not used in a pie.

This is a recipe my mother and her family used. In later years she used hamburger and if it was real fat she cut down on the amount of suet she put in.

*Gera Sawyer*
*Edgerton, Kan.*

## OLD FASHIONED MOLASSES CUSTARD PIE

1 cupful molasses
1 cupful sugar
1 1/2 tablespoonful flour

3 eggs
1 cupful milk
2 tablespoonfuls melted butter

Line a large pie plate with pastry, having a fluted edge. Combine the molasses, egg-yolks--slightly beaten, sugar, milk, flour and melted butter and fold in the egg whites last. Pour into the pastry lined pie plate and bake at 450º for 10 minutes to set the rim, then reduce the heat to 325º for 30 minutes. Two small pies may be made instead of one large one if desired.

Maxine Bowles
Knoxville, Ark.

## PLUM PUDDING

1 cup sugar
1 cup plums
1 teaspoon soda (dissolved in milk)
1 teaspoon cinnamon

1/2 tablespoon butter
3 tablespoons sour milk
1 teaspoon nutmeg
1 1/2 cups flour

Bake in slow oven for 30 minutes.

### SAUCE

1 cup fruit juice
1 tablespoon flour

1/2 cup sugar
1 tablespoon butter

Cook until it thickens and pour over pudding. This is another 1922 recipe from the Commerce Club.

Dorothy Beisinger
DeKalb, Mo.

## CHICKEN DUMPLINGS

Mix together and drop by small spoonfuls:

2 cups flour
1 teaspoon cream of tartar

1/2 teaspoon soda
1 cup sweet milk

Cover tightly and let cook for 15 minutes. This is another 1922 recipe from the Commerce Club.

Dorothy Beisinger
DeKalb, Mo.

## 1-2-3 PIE PASTRY

1 cup flour
3 tablespoons cold water

2 tablespoons lard
Pinch of salt

Mix flour and lard thoroughly. Add the water and mix well. Makes one double-crusted pie.
This was another 1922 recipe from the Commerce Club cookbook.

Dorothy Beisinger
DeKalb, Mo.

## SPICE CAKE

1 1/2 cup sugar
2 1/2 cups flour
3 eggs
1 teaspoon nutmeg
1 teaspoon cloves
1 teaspoon dark cocoa
1 tablespoon grape preserves

1/2 cup butter
3/4 cup sour milk
1 teaspoon soda
1 teaspoon cinnamon
1 teaspoon allspice
1 tablespoon dark jelly (any kind)

Bake for one hour in moderate oven. Top with caramel icing. Another 1922 recipe from the Commerce Club cookbook.

Dorothy Beisinger
DeKalb, Mo.

## APPLE NUT CAKE

1 1/2 cup oil
3 eggs
3 cups flour
1 teaspoon salt
3 cups peeled diced raw apples

2 cups sugar
1 teaspoon vanilla
1 teaspoon baking soda
1 cup chopped nuts
   (walnuts are best)

Mix together oil, sugar, eggs and vanilla. Add dry ingredients, nuts and apples. Bake in tube pan at 350° for 1 hour.

## SAUCE

Cook for 1 minute: 1 stick margarine, 1/2 cup brown sugar, 1/4 cup milk. Pour over cake immediately after removing from oven. Let set in pan until cool before removing. The cake and sauce are moist and keep a long time (if not eaten!) Can be served with topping.

Mrs. M.L Crouch<br>Richmond, Mo.

## TOMATO SOUP

Cook together for 5 minutes: 2 cups whole tomatoes and juice, 2 teaspoons sugar, 1 small onion (diced). Set aside.

Sauce: 2 cups milk, 4 tablespoons margarine, 4 tablespoons flour. Cook until mixture thickens and pour into first mixture.

Mrs. M.L. Crouch<br>Richmond, Mo.

## CORN CHOWDER (SOUP)

Brown 4 slices of bacon. Remove bacon and saute 1 medium chopped onion in grease. Add the following:

3 medium potatoes peeled and cubed   3 cups water
1 teaspoon salt                      1/4 teaspoon pepper

Cook all together until tender. Then add:

1 can tomato soup                    1 can whole corn, drained
(1) 12-ounce can condensed milk

Heat until hot. Do not boil.

Mrs. M.L. Crouch<br>Richmond, Mo.

## COTTAGE PUDDING

"This was always a favorite of my brother and sisters and myself. Whenever Mom would take the white pitcher down from the shelf, we knew we were going to have Cottage Pudding, or as we called it, "Cake 'N Sauce."

1 3/4 cups flour
1/2 teaspoon salt
3/4 cup sugar
1 cup milk

2 teaspoons baking powder
1/4 cup shortening
1 egg
1 teaspoon vanilla

Heat oven to 350º. Grease and flour an 8x8-inch pan. Mix flour, baking powder, and salt. Add egg, shortening, sugar and milk and vanilla. Beat until smooth. Pour into prepared pan and bake 25 to 30 minutes. Serve with the following sauce:

### Sauce for Cottage Pudding

1 cup sugar
2 cups water
1 teaspoon butter

2 tablespoons cornstarch
   (or use flour as my mother did)
1 teaspoon vanilla

Mix sugar and cornstarch (or flour) in saucepan. Stir in water and bring to boil, stirring constantly. Remove from heat and add vanilla and butter. Serve over generous serving of cake.

Donna Bartholow
Independence, Mo.

## MEAT SUBSTITUTE

Put one quart of water in pan, bring to a boil. Add one pint of corn meal. Cook until done, then add one can of chopped salmon. Let cool, then eat cool or fry.

Mrs. Charles V. Brokenicky
Smithville, Mo.

The following three recipes are from Florence Flanary, whose great-grandparents--all four sets of them--were among the many settlers who came from Ohio and Indiana to settle and create farms from the rolling fertile land that was to become Northwest Missouri.

## STEAM PUDDING

Take 1 cup sorghum and 1 egg. Mix together.
Add 1 teaspoon soda dissolved in 3/4 cup boiling water. Stir well. Add 2 cups flour and 1 teaspoon each cinnamon, cloves, allspice, nutmeg and salt. Add 1 cup, chopped and floured, each, dates, raisins, nuts. Mix together.
Steam over boiling water 1 to 1 1/2 hours until done when tested. Cool. Slice and serve with whipped topping or a favorite sauce.
"My mother made this pudding frequently during the winter from a recipe handed down in her family. She used a large aluminum pan with a fitted inner pan holding the pudding to steam over boiling water. Instead of this, pudding can be put in two greased and floured coffee cans and set over metal rack in pan of boiling water which can be tightly covered."

## HOLIDAY CAKE

"The following cake recipe was an invention of the eggless, butterless days of World War I, and continues to be a holiday favorite in our family. I bake it in an angelfood pan, ice it with white cooked frosting and top it with a wreath of candied fruit. It can be baked in a 9x13 loaf pan also."

2 cups water
2 cups sugar
1 cup lard
2 cups raisins
1 teaspoon salt

2 teaspoon cinnamon
2 teaspoons cloves
2 teaspoons nutmeg
2 teaspoons allspice

Mix together in casserole dish and bake in slow (325º) oven for 3 hours, stirring 3 times during first hour.

"The above recipe was used by my mother-in-law many years ago on Mondays, laundry day. It did not require much time and since she had the woodburning range going to heat the water for washing, it required no extra wood or effort."

Marilyn Holliman<br>Farragut, Iowa

## OLD TIME ITALIAN RECIPES

"Following are Italian recipes I have collected from old family recipes. My 85-year-old grandmother, Mary Grace Monteleone, cooks for our family daily. Like all strong Italian families who are close, this includes Sunday dinners for four generations at a time."

Mary Ann Guarino-McCartney<br>Kansas City, Mo.

## ITALIAN THUMBPRINT COOKIES

1/2 cup butter
1 cup sugar
1 teaspoon baking powder
1 teaspoon of orange, lemon, vanilla or almond extract
1/4 cup orange marmalade or 1/4 cup apricot jam
2/3 cup almonds, chopped

1 egg
2 cups sifted flour
2 tablespoons milk

Cream together sugar, egg, butter and flavored extract. Add baking powder, flour, and milk. Knead well, and pinch off small amounts of dough. Roll pieces in hand until they form balls. Roll balls in almonds and place on greased cookie sheets. With thumb, make imprint in center of each cookie ball. Fill each center with dollop of jam or marmalade. Bake at 400º for 12-15 minutes. Makes 2 dozen.

Mary Ann Guarino-McCartney<br>Kansas City, Mo.

## ITALIAN WINE STRIPS

2 cups flour
1/2 teaspoon baking powder
3 tablespoon sugar
1/4 cup butter

1/2 cup Italian red wine
2 cups oil
1/2 cup confectioners sugar
1 teaspoon ground cinnamon

Sift flour, measure, and resift with baking powder and sugar. Cut butter into flour with fingers until mixture resembles corn meal. Make a well in flour and pour wine into it. Knead dough until smooth--about 5 minutes. Wrap in waxed paper and set aside for 2 hours, but do not chill.

Heat oil in deep pan. Roll dough 1/4 inch thick, and cut into strips 1 inch wide and 4 inches long. Drop about 3 at a time in hot oil, and fry until golden brown. As they rise to oil surface, turn.

Remove from oil with slotted spoon and drain on paper towels. Combine confectioners sugar and cinnamon, and sprinkle over strips when they are cooled. Makes 2 1/2 dozen.

Mary Ann Guarino-McCartney
Kansas City, Mo.

## ITALIAN STUFFED MUSHROOMS

1 pound fresh mushrooms
1/2 cup bread crumbs
Salt and pepper
1/4 cup olive oil

1/4 cup grated Parmesan
  or Romano cheese
Pinch of oregano

Wash mushrooms with damp cloth and cut off stems. Place in shallow baking pan. Combine crumbs, cheese, seasonings and oregano, and sprinkle this mixture over mushrooms. Dribble oil over all, and bake in 350º oven till tender--about 15 minutes. Do not overcook. Serve at once. serves 4.

Mary Ann Guarino-McCartney
Kansas City, Mo.

## ROMAN PUNCH

1 cup water
Rind of 1/2 lemon
Juice of 3 lemons

2 cups sugar
Rind of 1/2 orange
Juice of 3 oranges

1/4 cup brandy  
(1) 26-ounce bottle chilled champagne  
2 tablespoons rum

    Boil sugar and water in saucepan over low heat about 10 minutes. Skim, and remove from stove. Add fruit rinds and juices, and cool to room temperature. Stir well, put in large jar, and chill 4 hours in refrigerator.

    Chill champagne and punch bowl. Strain syrup mixture through coarse mesh strainer into punch bowl. Add brandy, rum, and chilled champagne. Mix thoroughly, garnish with mint leaves or lemon slices, and serve in chilled punch cups. Serves 12.

Mary Ann Guarino-McCartney<br>Kansas City, Mo.

## ITALIAN HOT WINE PUNCH

Fifth of Italian red or burgundy wine  
Rind of 1 lemon  

1/2 cup sugar  

Rind of 1 orange  
Juice of 4 lemons or  
    1/4 teaspoon of ground cloves  
1 cinnamon stick

    Pour wine in stainless steel or enamel pan--don't use aluminum. Add cinnamon stick, orange rind, lemon rind, and sugar. Bring all to a boil, then reduce heat, and boil one minute more. Remove immediately from stove, and serve right away in cups or mugs. Serves 10 to 12 persons.

Mary Ann Guarino-McCartney<br>Kansas City, Mo.

## ITALIAN BROILED STEAK ROLLS

2 pounds tender beef or veal steak  
1 cup soft bread crumbs  
1 grated onion  

6 slices of diced white bread  
Salt and pepper to taste  

2 slightly beaten eggs  
4 tablespoons olive oil  

2 sprigs chopped parsley  
1/4 pound minced prosciutto  
1 tablespoon Romano  
    or Parmesan cheese  
1/2 cup flour  
1/4 pound Mozzarella cheese,  
    cut in 1-inch squares  
1 cup grated bread crumbs

    Cut steak into 8 slices, 3 inches wide, and flatten them to 1/4 inch to

1/2 inch thick. Have ten skewers ready. Combine parsley, onion, prosciutto, bread crumbs, grated cheese, and seasonings. Place part of this mixture on each of meat slices, and roll each slice up like a jelly roll. Place 1 meat roll, 1 square of Mozzarella, and 1 square of bread on a skewer. Roll each skewer lightly in flour, dip it in beaten eggs, and then roll in bread crumbs. Brush with olive oil, and place skewer on a preheated broiling rack. Broil slowly about 5 minutes, then turn each skewer, and broil other side 5 minutes. Serve hot on skewers. Serves 6.

Mary Ann Guarino-McCartney<br>Kansas City, Mo.

## SICILIAN SPAGHETTI CASSEROLE

2 tablespoons olive oill
1 pound sweet Italian sausage
   cut into 1-inch pieces
salt and pepper
1 pound cooked spaghetti
3 tablespoons gr. Parmesan cheese

1 small, chopped onion

1/4 pound sliced mushrooms
(1) No. 2 1/2 can tomatoes,
   or tomato puree
4 slices Mozzarella cheese

Brown sausage in the olive oil over medium heat about 10 minutes. Add the mushrooms, onions, and tomatoes. Add salt and pepper to taste, and cook uncovered over low heat for 1 hour, stirring occasionally. Prepare the spaghetti while the sauce is cooking. Place drained spaghetti in deep baking dish or pan, and cover with Mozzarella cheese. Pour sauce over cheese, and sprinkle with grated Parmesan cheese. Bake in oven for 350º until mozzarella cheese melts--about 10 minutes. Serve immediately. Serves 6.

Mary Ann Guarino-McCartney<br>Kansas City, Mo.

# Depression Days & Wartime
## 1930-1949

# Depression Days and Wartime, 1930-1949

## Gathering Greens

Betty Johnson Laverty has been writing for Discovery Publications for several years. Her "Parkville Nostalgia" columns have taken readers back to Depression Days in rural Platte County, where life was rich in the blessings of good neighbors and close families, if not in economic wealth. The next three stories are hers:

"Sitting and day-dreaming, my memoriy travels back to days long ago when I was a little blond-haired girl at home with my parents on Mace Road, a few miles out of Parkville, Mo. The late spring and early summer days seem, in my mind, always to have been warm and sunny, with a soft breeze ready to caress a cheek or ruffle a dress. In those days, little girls always wore dresses.

"The grass and trees were that special fresh spring green, and the sky a dazzling blue it always seems to be when we are quite young. And oh, how the birds did sing. Mom fed the birds from late fall to spring, so we always had flocks of them, and they rewarded her with hours of beautiful melody.

"Then there was the flower garden, carefully weeded by hand--no preservatives added--which was beginning to show its colors. If it was a very hot, dry summer, like we had in the 30's, the flowers were about the only things that got "watered." We had to pull our water up out of a well with a bucket and rope. No pulley or pump. Our drinking water was obtained this way, too. We were quite lucky, as our well was only about fourteen feet deep, and Dad could climb down on a ladder and "clean it out" whenever it was needed. The water was crystal clear, and quite cold. It was fed by an underground spring, and made a great "icebox" for keeping our milk and butter cold. Just set it in a bucket, tie a rope on the bail and lower the bucket so the bottom rested on the cold water.

"There was no electricity on Mace Road then. We had a dandy little ice box, but "town" was several miles away and we did not have a car. Dad walked to and from work at "the elevator" on Mill Street in Parkville. Of course, carrying a chunk of ice home was impractical. Once in awhile Noble Johnson (Dad's boss at the elevator) would give him a ride home, and we would be delighted to see a just slightly melted 10-pound ice chunk being taken from the car trunk.

"We had a little 5-acre place. Sometimes I think I can still smell the lilacs which bloomed on huge bushes standing in a row behind the house. My favorite spot on a hot summer day was the little fish pond, just a few feet away from the shaded screened-in porch on the east side of the house. A large shade tree was nearby, and the sparkling water of the pond seemed to always make this a cool, peaceful place to while away the hours. Tiny gleaming goldfish would dart in and out among the lily pads, and I was sure they were small bits of sun rays, come to earth to cool off.

"About once a week in the early growing season, we would go pick wild greens. It was a special happening for me, and I would jump for joy when Mom got out a paper sack or large pan, my little tin bucket and a paring knife. I never ate them--they looked and smelled too much like spinach--but I loved to pick them. It was many years before I realized why I always had my own container. Mom knew I'd pick any little plant that happened to catch my eye. She was careful to make sure I never saw her discard any of my "greens."

"I was going through some recipes my mother saved. Here is one of her favorite "greens" recipes:"

## TURNIP GREENS WITH CORNMEAL DUMPLINGS

Be sure to serve the pot likker with this!

4 ounces salt pork--leave rind on
1 pound turnips
1 teaspoon salt
1/2 cup flour
1 teaspoon sugar
3 tablespoons butter or margarine

3 quarts turnip greens
2 1/2 quarts boiling water
1 1/2 cups white cornmeal
1 teaspoon baking powder
1/2 teaspoon salt
1 beaten egg

Dice salt pork, wash and trim greens, pare and quarter turnips. In a large kettle, bring 2 1/2 quarts of water to a boil. Add salt pork, turnips, and 1 teaspoon of salt to boiling water. Simmer, covered, for 2 hours.

Remove one cup of likker and thoroughly stir together cornmeal, baking powder, sugar and 1/3 teaspoon of salt. Stir in the margarine and one cup pot likker, and add the egg. Spoon on rounded tablespoons onto the simmering greens.

Cover and simmer 30 minutes. Serves six.

Judy Johnson Kirtley, Parkville, Mo., says she's a big fan of Betty Laverty and reads her column in Discover Mid-America every month. Turns out she'd better--she's Betty's "little sister, Judy." Like Betty's recipes, Judy's are "old South" types, and date back to the early 1800's when their family settled in Arkansas. This next one goes well with their Mom's turnip greens and cornmeal dumplings recipe:

## ZESTY SAUCE FOR GREENS

1 1/2 teaspoon dry mustard
1/8 teaspoon pepper (black)
1 teaspoon onion, grated
1 teaspoon paprika
1/4 cup vinegar

1 teaspoon salt
1 tablespoon stuffed olives (chopped)
1 teaspoon sugar
1/3 cup olive oil or other light oil

Combine dry ingredients, stir in vinegar, add oil and remaining ingredients. Heat and serve with any greens. Also try on other vegetables, especially beets.
"This is an old southern recipe, originally from Arkansas."

## POTATO SALAD

"My mom was a champion potato salad maker, and was always urged to make it for every summer gathering. I am not sending a specific recipe here as I feel everyone has one available. I just want to suggest that one need not limit themselves to the old standard. Try adding one or two different ingredients now and then.
"Here is how Mom did it: first, she always used mashed potatoes. The flavors seemed to mingle better, without the risk of finding oneself with a mouthful of "not quite cooked" potato. Nowadays I find using instant mashed--minus the milk--works just fine. The amounts used of the following depends on your own taste."

*Chopped onions*--sweet--or if young onions are available, include the green blades, chopped. Adds color, too.
*Pimento pickles.* Chopped, sweet or dill. Try adding a bit of juice, also.
*Chopped celery, pepper (black).* Also a bit of green or red bell pepper.
*Salt,* including celery, garlic, etc.
*Fresh tomatoes*--diced.
*Vinegar,* salad mustard, horse-radish (ground) or sauce.
*Mayonnaise* or Miracle Whip.
*Several hard cooked eggs,* chopped, and a dash of paprika to top it off.

Mix sauce (mayonnaise, etc.) and spices together, THEN add to potatoes. Keeps them from being "gummy."
"Now to top the salad off, get fancy. Reserve one hard egg. Cut yolk in half (lengthwise) and place in the center of the salad, then slice the white and place around the yolk, forming a flower. Lightly sprinkle paprika or dill seed for color, and you have one pretty salad."

## WILTED LETTUCE

"My dad also had a large garden each summer, and when the leaf lettuce was in season, Mom served it wilted a couple of times a week. I have yet to be served any which was like hers, and I believe I have figured out why. See if you can spot the difference from most recipes."

About 1 lb. of leaf lettuce (or a large kettle-full.) I have heard head lettuce can be used, but have never tried it.
1 bunch radishes, red or white
1 bunch green onions
2 tablespoons bacon drippings
1/2 teaspoon salt
1 teaspoon sugar
1 tablespoon vinegar. More or less to taste. Cider vinegar works best.

"Wash lettuce well, shake off excess water as much as possible. Tear leaves into three or four pieces (each leaf), put all in large bowl. We used a crock or pottery type. Warm it in hot water just before using. Slice radishes and onions over top. Heat grease, vinegar, salt, pepper and sugar in heavy pan until it sizzles. Pour over lettuce, toss to mix well. Place a warm plate or lid over the top and allow it to wilt a few minutes. How long depends on how much you wish it to be wilted, so sneak a peek every few seconds.

"Crisp, crumbled bacon may be added immediately before serving, or place bacon pieces in small separate bowl and pass with lettuce, allowing everyone to add his own.

"Now the difference. All recipes I have seen say to serve immediately after adding hot mixture. Many prefer it this way, I realize. But we let it sit, covered for about 30 to 40 seconds."

## TOMATO-ONION SALAD

"This was another of Mom's "specials." Once more I cannot give exact amounts--just depends on how you like it. Only fresh garden tomatoes and sweet onions should be used.

"Generally, four large tomatoes to one large onion is the right proportion. Chill the tomatoes and the onion. If tomato skins are a bit tough, peel tomatoes and remove most of the seeds. Cube onion and tomatoes in equal sizes, mix, add salt, pepper and a bit of sugar to taste. Serve with meal. Goes well with cottage cheese, too. Salad should not be salty or sweet. Use just enough of each to enhance the tomato flavor. Use the old "add and taste" method."

Judy Johnson Kirtley<br>Parkville, Mo.

# Walking to School

**B**etty Laverty was one of the army of farm kids who lived through the Depression Days and emerged with poignant memories--and a sense of humor. Walking to school meant facing some challenges that modern students in urban communities may never know. Her story:

"As a child I walked to school, as did all the rest of the students of Graden School. It was a long trip for a first grader, all alone, trudging those four miles of country roads in all kinds of weather. However, we didn't think too much about it then, as it was all we knew. Sometimes we walked as a group, and when we did there was quite a bunch of us. The Samborksi kids lived at the top of the hill, so they started first, then I joined them. Pauline and I walked along together. Sometimes Tony joined us, but usually he and his brother, Baby Edward, ran on ahead. I never did figure out why Eddie ran all the way to and from school, but there he'd go, with his lunch wrapped in paper, tucked tightly under his arm, his short little legs churning. He would soon disappear over the top of a hill.

The older Samborski girls walked together, keeping a keen eye on the younger ones. The Samborski family made their living farming. They grew vegetables in the summer and took them to market to sell. All the kids worked in the fields and sometimes had to miss school in order to get the planting done. No telephone call, written excuse, or conference with the teacher was needed. Having grown up in a farming community, the teacher understood that if the kids were absent it was because they had to stay home and work in order to insure the family's survival. Kids did not skip school then--it was much preferred to staying home and working.

"Next we picked up the Gresham kids, then the Reynolds' and finally Mary Ellen and Jimmie Reid. This totaled 16 kids in all. However, in winter we seldom all walked together, as it was too cold to stand and wait for anyone. Besides, no one wanted to be late for school It just wasn't done.

"Walking to school was serious business, but walking

home was quite another story. There was time to stop at Thompson's creek and watch the water bubble over the rocks. And in the warm days of fall, this was a good time to pull off the still unfamiliar shoes and rest hot, tired feet in the cool water. In the spring, there was time to gather wild flowers, then hurry home before they wilted. Of course some kids had to hurry on to help with the chores. And, as always, Baby Edward ran!

"We usually cut through Ernie Thompson's as it saved us about a mile's walking distance compared with following the road. The cut-through was one long hill down, covered with trees, rock and weeds, then a meadow through which a small creek ran (it could be a raging torrent), up a hill such as we had just descended, across a pasture area, through a large metal gate, down a driveway, and we were at the highway. Then school was only a hop and skip away.

"We kids thought the Thompsons were quite nice. In all those years they never once, to my knowledge, said a word about all the kids who crossed their fields and tromped down their blacktop driveway with muddy shoes. Of course, we were always careful to close the gate, which could be quite a task for a little girl (or boy) who looked like she never drank her Ovaltine (remember those ads?) and never picked a flower from the yard.

"I must tell you about an experience I had with that gate. I was a second grader, and this day I was alone. Mom did not like me to cut through when I was by myself, but it wass especially cold, and I was in a hurry to reach the warmth of school, so I took a chance she'd not find out.

"Arriving at the large metal gate--it must have been about eight feet long--I couldn't get it unfastened due to the heavy ice coating. Finding a rock, I gave the latch a good wallop, breaking the ice coating away. This allowed me to lift the latch, and the gate began its slow, easy swing open. Being a "farm kid," I knew not to swing on gates, especially someone else's, as it tended to pull the hinges out of the post. But for some reason, I grasped the gate with both hands, stuck my nearly frozen toes through the wire mesh, and rode the gate its full lazy arch until it came to a stop. Now...instead of hopping off, pulling it closed and going on my way, I did something I have never been able to explain. Like it had a mind of its own, my tongue snaked right out of my mouth and stuck to that gate.

"Horror of horrors--I could not pull it back! It was stuck firmly to the cold metal. Picture my plight, and what a picture it must have been--this little girl clinging desperately to the gate, praying she would not slip, lest she have her tongue yanked clear out of her mouth--yet wanting to get free before anyone saw her. In my mind I could see the Thompsons finding me with my tongue stuck to their gate, and finally a whole hoard of kids from school would gather, and laugh--and laugh. I'd never be able to go to school again. If I did, they'd call me something like "Old Gate Tongue" or worse. Besides, I probably woudn't have a tongue at all, for if it didn't get yanked out it would surely freeze, swell, and turn black. Then they'd have to cut it off.

"I wondered if Mom and Dad would be made at me or feel sorry for me. I knew all the kids would hate me, because the Thompsons would *never* let us cut through again. But of course, no one would want to play with me again--after losing my tongue.

"It seemed I clung there forever. But in fact it was only a few seconds, for to my great relief the body heat of my tongue warmed the metal to the point of release, and I was able to free myself. Only a small piece of skin from my tongue was missing. I closed the gate and hurried on my way.

"No one knew of this experience until many years later, when one of my little girls got her tongue stuck to an ice cube tray. My memory flashed back as I saw the look of terror in her big blue eyes. In spite of myself I had to laugh. Of course this called for an explanation, as non of my three daughters saw anything funny about the situation. So I gathered them together and told them the "gate" story. They're grown now, but they can't help bursting into peels of laughter whenever they recall the story of their mom getting her tongue stuck to the gate."

## Another Thanksgiving Remembered

As the days dwindle down, and Thanksgiving time comes, today's homemakers check out the frozen turkeys and squeeze the family budget for extra treats for the big dinner. In the Depression-poor 30's, it took a lot more squeezing. Betty Laverty remembers:

"During the 1930's we were just thankful for whatever we had to eat. As we raised chickens, there usually was an old hen to roast, dried bread for the dressing, potatoes, home canned tomatoes and applesauce. And if there was pie it was pumpkin or real mincemeat--made with suet.

"But at Graden School, our teacher, Miss Miller (now Velma Loncar) would put up a large picture of a "typical" family gathered around the table, awaiting a slice of the large, golden-brown turkey. We made paper turkeys to decorate the blackboard and windows, and studied how the first Thanksgiving came to be. How I longed for a "typical" turkey feast.

"Finally it happened--almost.

My Grandpa Johnson had been given a big fat goose in payment for a job he had done, and Grandma Alice invited us to join them for dinner. Mom assured me a goose was quite like a turkey, and would make a fine showing.

"When the big day arrived, we loaded the food we were to provide and headed to Grandpa's. The smells of onions, sage, and roasting fowl greeted us when we arrived. Soon the men settled down in the living room, and Bernice and I were given the job of setting the table. We took great pains to get everything just right, not forgetting the dill pickles Mom had brought. Then we were told to "scat" while Mom and Grandma completed the final touches of the meal.

"Finally, the moment arrived. The bird, complete with dressing, was lifted out of the oven. What a sight it made, done to a perfect golden brown. What a Thanksgiving we were to have. We took our places at the table. Grandpa "returned thanks," and Dad was asked to do the carving.

"He had not gotten very far along with his job when a strange look crossed his face. In an embarassed tone of voice he asked, "Alice, did you remember to remove the grease sack from this goose?"

"Well the dogs had a fine Thanksgiving, complete with goose and dressing. And really, the bologna and cheese sandwiches, along with the rest of the trimmings, were quite tasty.

"At least for a few minutes, we had a perfect Thanksgiving Day meal!"

## Judy's Special Christmas

All of us have memories of special Christmases as children. Betty Laverty's sister, Judy, described a poignant childhood Christmas and Betty included the story in her "Parkville Nostalgia" article in the December, 1987 issue of *Discover KC*. It seems a fitting conclusion to our feature section. The year was 1949.

"Every Tuesday after school, I went to Brownie Scouts at the Parkville Community Church. Then I walked downtown lto wiat for Dad to give me a ride home. He always met the 5:30 Greyhound bus in Parkville, as my sister Betty rode it to and from work in Kansas City. Then the three of us rode on home together.

"I liked these times. It gave me the chance to look around, and my favorite place to look was Miller's General Store. Miller's contained all sorts of interesting things, but most of all it was the dolls I loved to look at. Alma always made me feel welcome, and I was free to snoop at my heart's content.

"One Tuesday in mid-November I hurried in, quickly closing the door behind me to keep out the cold, blustery wind. "Hello, Judy," Alma greeted me. "Think winter's here?" I turned to answer her, then stopped short, my manners completely forgotten. There on a shelf behind the counter sat the prettiest doll I had ever seen. She was a baby doll, all dressed in fresh white cotton, including the bonnet and booties, trimmed with tiny pink rosebuds. It was love at first sight between Sweetie Pie and me.

"Following my gaze, Alma smiled. "I knew you'd be glad to see I have my Christmas dolls in. See one you like?" Still without speaking, I moved forward to stand directly in front of *her*. "Oh, I love you," I whispered. I had failed to even notice the other new dolls, and had completely forgotten Alma was in the room. She finally reached me when she told me the Greyhound was in.

"I shook free of my trance, rushed out the door and down the street. Betty just must see Sweetie, she had to. Betty was 13 years older than me, and always took the time to make me feel "special," buying me pretty dresses in Kansas City, brushing my hair, and most of all, listening to me. Hand in hand, we hurried up the street to Miller's.

"See, there she is," I exclaimed. "Oh, Judy, you're right, she really is very special," she agreed. Alma took Sweetie Pie from the shelf and placed her in my arms. "I think Judy's quite taken with this doll," she said. "Don't think it will hurt, just this once, to let her hold it for a minute." Alma was a good person, and a smart business woman.

"Betty, I must have her, I just must," I said, as Alma put her back on the shelf. "I don't know, honey, she's quite expensive," my sister replied. "But Santa could bring her, couldn't he?" I asked. "Well, that's true," Betty said, "we'll just have to wait and see." And with this she hurried me out the door as Dad was waiting for us.

"In the days that followed I could hardly keep my mind off Sweetie Pie, and every chance I had I paid her a visit. One Saturday Santa came to Miller's and Dad took me to town especially to see him. I liked him immediately. Besides, he reminded me of Leland Francis-- even had a Masonic ring just like Leland's. I told him all about Sweetie, and even pointed her out to him. "Well, now, Judy, i think I can just about promise you you'll have your doll on Christmas morning," Santa chuckled, and smiled at Dad. As we left the store Dad muttered something like "old devil shouldn't go around promising things." But I didn't pay any attention; he was always teasing me.

"The next time I stopped into visit "my" doll, my world fell apart! I just stood and stared at an empty spot where she had been. Had Alma moved her? But she just shook her head and told me someone had bought her. But Santa had promised!

After that, I just couldn't get into the tree decorating and other Christmas preparations. If I wasn't going to have my doll, I wasn't much interested in anything else. My family tried in vain to cheer me up--even my big brother Ralph, and he hated my dolls, always calling them "brats" and such--felt sorry for me. and it hurt that Santa had lied to me. I just never, never, would believe him again. Dad was right, after all!

"That Christmas Eve I went right to bed, and went to sleep. Usually Ralph and I could never get to sleep on this night until late.

"Later that night, or very early in the morning, I woke briefly and heard voices in the living room. I even thought I heard a baby's cry, but I knew it wasn't Sweetie, so just rolled over and went back to sleep--wasn't even tempted to peek.

"I was awakened later by Ralph's anxious voice as he shook my shoulder. "Judy, wake up! come see what Santa has left!" In spite of myself I hurried to follow him into the living room, then stopped short in front of the tree as it stood winking and blinking at me.

"For an instant I stood and couldn't even move, or believe my eyes. Th,en I slowly knelt by the tree, taking my gift in my arms. "Sweetie Pie, oh Sweetie Pie, I love you!" I held her soft little body close, and felt a couple of tears escape from my eyes and roll down my cheeks. From somewhere far away I heard Betty's voice creep into my world as she placed a loving hand on my shoulder. "See, Judy, Santa **didn't lie after all!**"

"I am grown now, even a grandmother. But Sweetie was--and always will be--very special to me. She sits in her own little white wicker rocker in my room, and is not only a beloved friend from my youth, but a constant reminder to me not ever to lose faith in those things I believe in."

## GRANNY'S CARROT COOKIES

1 cup shortening
1 cup carrots, mashed
2 cups flour
1/2 teaspoon salt

3/4 cup sugar
1 egg, unbeaten
2 teaspoons baking powder
1 teaspoon vanilla

FROSTING:

Juice of 1/2 orange
1 tablespoon margarine

1 cup powdered sugar

Cream shortening and 3/4 cup sugar. Add carrots, egg, and vanilla. Sift flour, baking powder, and salt. Drop by teaspoonful on greased cookie sheet. Bake at 350° from 8 to 10 minutes.
To frost warm cookies, mix orange juice, margarine and powdered sugar.

Debbie Sieger
Clinton, Mo.

## HOMEMADE MAYONNAISE

1 1/2 cup sugar
1 1/2 cup cream or half & half, soured
1 cup vinegar

1 teaspoon salt
4 eggs
1 heaping tablespoon flour
1 tablespoon butter (set aside)

Using a double boiler, combine all ingredients and cook until thickened. After lifting from heat, add butter. Cool. Pour into jars and refrigerate. Makes 2 quarts.
Our family enjoys this on sliced bananas with chopped pecans, ham salad, pimiento cheese, macaroni salad or fruit salad.

Debbie Sieger
Clinton, Mo.

## BAKED APPLES

6 apples
1 cup water
1/2 cup red hots

1/2 cup sugar
3 tablespoons butter
1/2 teaspoon cinnamon

Place halved and peeled apples (cut side up) in dish. Mix remaining ingredients over heat until red hots and sugar are dissolved. Bake at 350º until soft. About halfway through baking time, turn apples so rounded side is up.

Debbie Sieger
Clinton, Mo.

## COUNTRY BROWN BREAD

1 1/2 cup sour milk
Pinch of salt
3/4 cup sugar
1 1/2 cup white flour

1 1/2 tablespoon sorghum
1 1/2 cup all-bran
1 1/2 teaspoon baking soda

Combine milk, sugar, sorghum, baking soda and salt. Stir well. Add flour and all-bran. Beat well. (May add nuts or raisins if desired.)
Bake in well-greased tin cans 45 minutes at 325º. Cover the first 15 minutes of baking time. Fill cans only half full.
Makes three 17-oz. cans of bread. Great with honey butter.

Debbie Sieger
Clinton, Mo.

## FRESH CHOPPED APPLE CAKE

1 cup brown sugar
2 eggs
1 cup sweet milk
1 teaspoon cinnamon
1/4 teaspoon nutmeg

1 cup white sugar
1/2 cup shortening
2 1/2 cup flour   1/2 teaspoon salt
1 teaspoon soda
2 cups chopped apples

Cream shortening and sugar. Add eggs. Mix dry ingredients, add to sugar mixture. Add milk and chopped apples. Place in well greased and floured 9x13 pan. Top with:

2/3 cup brown sugar
1/2 cup chopped nuts

1/2 cup margarine
1 cup coconut

Mix well and drop by small spoonfuls on top of unbaked cake. Bake at 350º for 45 minutes.

Ellen Barnett
Kearney, Mo.

## VINEGAR EGG PASTRY

3 cups flour
1/2 cup cold water
1 large egg

1 teaspoon salt
1 cup shortening
1 teaspoon vinegar

Sift salt with flour and cut in shortening. Beat egg, adding water and vinegar. Add liquid to flour mixture. Mix until blended.

Place pastry on lightly floured board. Knead 20 strokes. Then roll pastry out. This will make three crusts. Keeps well in refrigerator for two weeks or more. This is a never-fail pie crust.

Ellen Barnett
Kearney, Mo.

## PEPPERONI SAUSAGE

5 pounds lean ground beef
5 teaspoon mustard seed
4 teaspoon garlic salt
1/2 teaspoon crushed red pepper Pepper

1 teaspoon fennel seed
1/2 teaspoon anise seed
2 teaspoon hickory smoked salt
5 teaspoon coarse ground black
5 rounded teaspoons tenderizing salt

Day 1: Mix in bowl and refrigerate.
Day 2: Mix by hand and refrigerate.
Day 3: Repeat Day 2.
Day 4: Form 5 rolls and bake on broiler rack in bottom of oven 8 hours at 140º. Turn every 2 hours.

Ellen Barnett
Kearney, Mo.

## DEPRESSION DAYS CAKE

1 1/2 cup self-rising flour
1 teaspoon soda
1 tablespoon vinegar

1 cup sugar
2 heaping tablespoons cocoa
1/3 cup melted grease

2 teaspoons vanilla                    1 cup butter

Mix together the dry ingredients, poke three holes in this and put the vinegar in one, grease in the second and vanilla in the third. Pour in the water over all, stir well and bake until brown--about 35 to 40 minutes at 350º.

Frances Uttinger<br>Kansas City, Kan.

## NO FAIL DIVINITY

2 cups sugar                           1/2 cup water
1 (7 oz.) jar marshmallow creme        Pinch of salt

Boil sugar, water and salt to hard ball stage. Put marshmallow creme in a large bowl and stir in hot syrup, stirring by hand.
Continue to stir until mixture is slightly stiff and holds a peak. Fold in vanilla and drop by teaspoon onto waxed paper.
Nuts may be stirred in at the last.
I never could make divinity until I tried this recipe. Fool proof!

Frances Uttinger<br>Kansas City, Kan.

## TENDER HOMEMADE NOODLES

1 cup flour                            1 egg
1/2 teaspoon salt                      1 tablespoon butter or oleo
1/2 teaspoon baking powder             2 tablespoons milk

Sift flour into bowl, shape a well in center and add remaining ingredients. Mix until a stiff dough is formed. May have to use your fingers.
Roll out very thin on floured board and let stand 15 to 20 minutes. Roll up dough and slice into strips. Drop into chicken broth or beef bouillion. Cook 10 minutes or until tender. May make several batches and dry to have on hand or freeze them right away.

Frances Uttinger<br>Kansas City, Kan.

## JESSIE'S REFRIGERATOR FRUIT CAKE

*Do Not Cook*

2 packages chopped pecans (16 oz.)
1 box yellow raisins
Large package coconut
1 box vanilla wafers (crushed)
2 bottles maraschino cherries chopped--save juice
1 small can sweetened condensed milk--add to cherry juice

Mix all together. Pour in flat pan, press down with a wet spoon, then put in refrigerator.

Frances Uttinger
Kansas City, Kan.

## CARROT SALAD

2 cups grated carrots
2 tablespoons lemon juice
3 tablespoons sugar

Place grated carrots in bowl. Sprinkle on sugar and pour lemon juice over the sugar. Do not stir.

Place in refrigerator for at least two hours--longer if possible. Stir and serve. The longer it sets, the better it is.

NOTE: do not salt, as the salt will wilt the carrots and ruin the crispness.

My mother got this recipe from a State Home Economics session she attended in the late 1930's.

Gera Sawyer
Edgerton, Kan.

## TEXAS WHITE CAKE

1/2 cup shortening
1 3/4 cup flour
2 cups sugar
1 3/4 cup buttermilk
1/2 teaspoon almond extract
2 egg whites
1 teaspoon baking soda
1/2 teaspoon salt
1 teaspoon vanilla

Cream shortening and sugar, add unbeaten egg whites. Mix together flour, soda and salt. Add to creamed mixture, alternating with buttermilk and flavorings. Pour into greased and floured 17x11 cookie sheet. Bake 20 minutes at 350º.

ICING:

1 stick butter
1 box powdered sugar
1 1/2 cup coconut
1/3 cup buttermilk
1 teaspoon vanilla
1/2 cup chopped pecans.

Melt butter, add buttermilk and rest of ingredients. Mix well and spread on cake after it cools 5 minutes.

Faye Ann Roberts
Kansas City, Mo.

## PINEAPPLE SHEET CAKE

2 eggs
2 cups flour
1 teaspoon soda
1/2 to 1 cup chopped pecans
2 cups sugar
1 teaspoon vanilla
(1) No. 2 can crushed pineapple
Do not drain

Mix. Bake at 350º for 45 minutes. Bake in a 9x13 pan.

FROSTING

4 oz. cream cheese
2 cups powdered sugar
1/2 cup chopped walnuts
1/2 cup butter
1 teaspoon vanilla

## ORANGE SLICE CAKE

1 cup butter
4 eggs
3 1/2 cups flour
1/2 cup buttermilk
1 can flake coconut
1 tablespoon orange rind
2 cups sugar
1 teaspoon soda
2 cups pecans
1 lb. orange slices (cut up)
1/2 lb. dates (cut up)

Roll fruit in 1/2 cup flour. Mix all together and bake in tube pan at 225º for 3 hours or until straw comes out clean. Bring 1 cup orange juice with 1/2 cup sugar to boil. Pour over cake while hot and leave in pan overnight until cold.

Faye Ann Roberts<br>Kansas City, Mo.

## SPINACH AND SPAGHETTI

2 packages frozen chopped spinach
2 tablespoons minced onion
2 beaten eggs
Cook and drain.

1 1/2 lb. Monterey Jack cheese, grated
8 ounces sour cream
Spaghetti, 8 oz. (uncooked weight)

Cook spinach as directed on package and press in a strainer to remove excess water. Combine with remaining ingredients in a greased casserole. Bake at 350º for 35 minutes or until bubbly.

Faye Ann Roberts<br>Kansas City, Mo.

## PINK ARCTIC FREEZE

6 ounces cream cheese
2 tablespoons sugar
1 c pineapple tidbits drained or crushed
1 cup whipping cream, whipped

2 tablespoons mayonnaise
(1) 1 lb. can whole cranberry sauce
1/2 cup chopped walnuts

Soften cheese; blend in mayonnaise and sugar. Add fruits and nuts, fold in whipped cream. Pour into loaf pan. Freeze firm for six hours or overnight.

To serve, let stand at room temperature about 15 minutes. Turn out onto lettuce; slice.

Faye Ann Roberts<br>Kansas City, Mo.

## OATMEAL COOKIES

1 cup raisins
1 cup sugar
1/2 cup dates
2 cups flour

1 cup shortening
3 beaten eggs
1 cup nuts
1 teaspoon salt and soda

1 teaspoon cinnamon
1/2 teaspoon cloves

1/2 teaspoon allspice
2 cups rolled oats

Cover raisins with water. Boil 5 minutes to liquefy raisins.
Cream shortening, sugar and eggs. Mix oats, flour, soda and salt with 6 teaspoons of the raisin liquid. Add 1/2 cup of dates and 1 cup chopped nuts.
Bake at 375º for about 12 minutes.

Faye Ann Roberts<br>Kansas City, Mo.

## AUNTY'S COLE SLAW

4 cups chopped cabbage
2 tablespoons vinegar

4 tablespoons sugar

Place cabbage in bowl, sprinkle with sugar, drizzle vinegar over all.
Do not add salt. This wilts the cabbage. Place in refrigerator for several hours. Stir well and serve. Keeps well.
This was given to me in the 1940's by a cousin who was quite a bit older than I.

Gera Sawyer<br>Edgerton, Kan.

## HOOSIER CHOCOLATE SHEET CAKE

2 cups flour
1 teaspoons baking powder
1 cup butter
4 tablespoons cocoa
2 eggs

2 cups sugar
1/2 teaspoon salt
1 cup water
1/2 cup buttermilk
1 teaspoon vanilla

Sift together flour, sugar, soda and salt. Melt butter, water and cocoa, bring to a boil and pour over dry ingredients. Add buttermilk, eggs and vanilla. This is very thin batter.
Pour into greased 16x10 jelly roll pan and bake for 20-25 minutes in a 350º oven.

While still hot, spread with following:

1/2 cup butter
1 cup nuts
1 pound confectionary sugar

6 tablespoons buttermilk
4 tablespoons cocoa
1 teaspoon vanilla

Melt butter. Add buttermilk and cocoa and boil one minute. Add sugar, nuts and vanilla and spread on cake.

Faye Ann Roberts
Kansas City, Mo.

## JUDY'S CHOCOLATE PIE

8 ounce chocolate bar
3 teaspoons whipped topping

17 large marshmallows
1/2 cup milk

Melt all of the above in double boiler. Put in graham cracker crust and top with whipped topping.

Faye Ann Roberts
Kansas City, Mo.

## PLUM CAKE

1 cup salad oil
3 eggs
1 teaspoon each allspice, nutmeg, cinnamon
1 cup pecans chopped

2 cups sugar
2 cups self rising flour

1 jar, junior size baby plums

Cream oil and sugar. Add eggs, alternating with flour, spices included. Add plums and nuts.
Bake in tube pan 1 hour at 325º.

Faye Ann Roberts
Kansas City, Mo.

## CRANBERRY SALAD

(1) one-lb. bag of cranberries
1 pint whipping cream
2 cups chopped nuts

2 cups sugar
1 cup mini marshmallows

Chop (I use my food processor) cranberries and add sugar. Mix well.

Allow this to sit overnight (in refrigerator).

Next day, beat whipping cream until it reaches a very stiff consistency. Fold it into the cranberry mixture along with the marshmallows and nuts. Refrigerate 4 hours before serving.

Kathy Osborne<br>Sedalia, Mo.

## SPECIAL BAKED CHICKEN

(1) 3-oz. package sliced dried beef
3 large chicken breasts, boned and skinned
1 can mushroom soup
6 slices bacon                    1 cup dairy sour cream

Run cold water over dried beef. Drain and arrange on bottom of 12x8x2 dish.*

Place halved chicken breasts over beef. Top each with bacon.

Sprinkle a little rosemary on top. Bake uncovered at 350º for 30 minutes.

(*Or use slow cooker several hours.)

Kathy Osborne<br>Sedalia, Mo.

## HERSHEY CHOCOLATE CAKE

Cream together:

2 sticks butter                    2 cups sugar

Melt in double boiler:

1 can Hershey chocolate syrup      (1) 8 7/8 oz. size candy bar, Hershey
Cool and add to creamed mixture.

Add alternately:

1 cup buttermilk                   1/2 teaspoon soda
2 1/2 cup flour                    1 cup chopped pecans

Bake in tube pan for 2 hours at 300º. Cool and wrap in aluminum foil.

Keep for 3-4 days before cutting.

Faye Ann Roberts
Kansas City, Mo.

## RAW APPLE CAKE

1 cup oil
2 eggs
3 cups flour
1 teaspoon salt
1 cup nuts, chopped

2 cups sugar
3 cups peeled apples (diced)
1 teaspoon soda
3 teaspoons cinnamon

Mix oil, sugar, eggs. Add apples, blend. Sift dry ingredients, add to apple mixture and add nuts. Put in greased bundt pan. Bake at 350º for one hour.

GLAZE:

1/4 cup evaporated milk
1 stick butter

1 cup brown sugar

Boil in pan for two minutes. Cool, spoon over cake.

Faye Ann Roberts
Kansas City, Mo.

## GREEN TOMATO MINCE PIE

1 1/2 quarts green tomatoes
2 cupfuls chopped tart apples
1 lb. raisins or mixed chopped fruit
3 cupfuls medium brown sugar

Salt
1 chopped orange
2 Tbls. mixed ground spices

Slice the tomatoes very thin, sprinkle with salt and let stand overnight. Then drain and chop very fine. Add the chopped tart apples and chopped orange. Simmer two hours. Add the brown sugar, raisins or mixed chopped fruit, and mixed ground spices, and simmer one hour. This amount is sufficient for two good-sized pies.

Cool the filling before putting into the pie crust. Put on top crust and bake at 450º for 30 minutes.

Maxine Bowles
Knoxville, Ark.

## SPOOK SALAD

Lettuce
Apples
Mayonnaise
Cloves

Celery
Red Grapes
Peaches
Pimiento

Make a nest of lettuce leaves, or shredded lettuce. On this, place a salad made of celery, apples and red grapes, allowing for each serving about two tablespoons each of chopped celery and apples, and six grapes cut in half and blended together with mayonnaise.

On this salad place a half peach with the round side up. Insert two whole cloves with the heads for the eyes. Place another with the large end down for the nose and a narrow strip of pimiento for the mouth.

Maxine Bowles
Knoxville, Ark.

## DEEP DISH PIZZA

1 1/2 pound ground beef
6 ounce can tomato paste
  or pizza sauce for more spice
1 to 1 1/2 teaspoons oregano
1 1/4 teaspoons salt
2 ounce can (1/4 cup) Mozzarella cheese, shredded

1/2 cup chopped onion or
  2 tablespoons instant onion
2 tablespoons gr. Parmesan cheese
1/2 teaspoon pepper
1 can (10) biscuits

Heat oven to 350°. Grease nine-inch pie pan. Brown ground beef and onion; drain. Stir in tomato paste, one tablespoon Parmesan cheese, oregano, salt and pepper. Simmer while preparing crust.

Separate dough into ten biscuits over bottom and up sides to form crust. Spoon hot mixture into crust. Place mushrooms over meat mixture. Arrange tomato slices over pie; sprinkle with Mozzarella cheese and remaining Parmesan cheese.

Bake in 350° oven for 20 to 25 minutes. Cool five minutes before serving. Serves five to six.

I double this for a 9x13 inch cake pan.

Kathleen Potter
Savannah, Mo.

## BEEF CRUST PIZZA
*(For diabetic or heart patient)*

6 servings (yield: a 10-inch pizza). 1 serving: 1/6 of pizza.

### CRUST:
1 pound lean ground beef
1 medium egg, beaten
1/2 teaspoon garlic powder
1/4 teaspoon pepper
1 slice bread, finely crumbled
2 teaspoons onion powder
1 teaspoon salt
1/2 teaspoon oregano

### FILLING:
1 (6 oz.) can tomato paste
1 1/4 cup thinly sliced sweet green pepper
1 (4 oz.) can mushroom pieces, drained
4 ounces shredded mozzarella cheese
1/2 teaspoon basil

Preheat oven to 375°. Combine all crust ingredients and mix thoroughly. Press into a 10-inch pie pan to form a crust. Bake 10 minutes. Leaving crust in the pie pan, drain off all liquid fat. Spread tomato paste on bottom and sides of crust, then sprinkle with basil. Spread even layers of sliced green peppers, mushrooms, and cheese. Bake 15 minutes.

Serve immediately, cutting pizza into six equal wedges.

For low-sodium diets, omit salt. Use unsalted tomato paste and unsalted canned mushrooms.

Kathleen Potter
Savannah, Mo.

## NO SUGAR FRENCH BREAD
*(from Crystal Adkins)*

2 1/2 cups warm water (105° to 115°)
1 tablespoon salt
7 cups unsifted flour
1 egg white
1 tablespoon cold water
2 packages or cakes yeast, active dry or compressed
1 tablespoon melted margarine
Cornmeal

Measure cold water into large warm mixing bowl. Sprinkle or crumble in yeast; stir until dissolved. Add salt and margarine. Add flour and stir until well blended. (Dough will be sticky.)

Place in greased bowl, turning to grease top. Cover; let rise in warm place, **free from draft, until double in bulk**, about one hour.

Turn dough out onto lightly floured board. Divide into two equal portions. Roll each into an oblong 15x10 inches. Beginning at wide side, roll up tightly towards you; seal edges by pinching together. Taper ends by rolling gently back and forth.

Place loaves on greased baking sheets sprinkled with cornmeal. Cover; let rise in warm place free from draft, until doubled in bulk, about one hour. With razor, make four diagonal cuts on top of each loaf.

Bake in hot oven (450º) for 25 minutes. Remove from oven and brush with egg white mixed with cold water. Return to oven. Bake five minutes longer.

Kathleen Potter<br>Savannah, Mo.

## WINDOW WASHING SOLUTION

1 pint alcohol
2 tablespoon ammonia

2 tablespoon liquid soap
About 8 pints of water

SMALL AMOUNT:

1/4 cup alcohol
3/4 teaspoon ammonia

3/4 teaspoon liquid soap
About 2 cups water
(enough to fill a pint)

Combine in container the alcohol, soap and ammonia. Add enough water to fill gallon container. May add a few drops of coloring to make it distinguishable from other fluids.

Kathleen Potter<br>Savannah, Mo.

## ONION SOUP AU GRATIN

3 cups meat broth
1/2 teaspoon salt
2 tablespoons cold water

6 medium sized onions, chopped
4 tablespoons flour
Pepper, toast, Parmesan cheese

Cook the chopped onions in a small amount of water until tender. Add two tablespoons of fat from the meat broth or the same quantity of butter and let the onions cook down in this until they are yellow. Mix them with the meat broth and salt and thicken with the flour and cold water, which has been blended.

Pour soup into bowls or soup plates. Place on top a round or slice of toasted bread and sprinkle grated cheese over the bread and soup. Serve at once.

Frances Uttinger<br>Kansas City, Kan.

## LEMON PIE

Juice and rind of 1 lemon
1 heaping tablespoon cornstarch
2 eggs

1 cup sugar
1 teaspoon butter

Mix sugar and cornstarch. Add well-beaten egg yolks, lemon and butter. When well-mixed, pour all into one pot of boiling water and let cook a few minutes.

Beat whites of two eggs to stiff froth. Add two teaspoons sugar and spread over filling. Brown slightly.

Frances Uttinger<br>Kansas City, Kan.

## BURNT SUGAR ANGEL FOOD CAKE

One pint cake flour, 1 1/3 pint sugar. Sift 12 times together. Add 15 egg whites, 1/4 teaspoon salt.

Beat until egg whites will not slide when crock is turned upside down. When partly beaten, add scant 3/4 teaspoon cream tartar. Finish beating. Add 1 teaspoon vanilla, 2 tablespoons burnt sugar.

Stir flour and sugar in very slowly. Bake in ungreased pan 1 1/4 hours in slow oven.

### ICING

Use 2 egg whites, 5 tablespoons sugar. Add 1 teaspoon burnt sugar flavoring. Cook over hot water in double boiler. Beat with egg, beaten until stiff. Remove, add 1 teaspoon vanilla, frost cake.

Frances Uttinger<br>Kansas City, Kan.

## FLOUR PUDDING

Work three ounces of fine flour and three ounces of butter into a paste. Put this into a pint of good boiling milk, and stew it on the fire until the mass separates from the stew pan.

Then take if off and let it cool a little, stirring into the mass the yolks of six eggs, two ounces of sugar, the grated peel of one lemon, and finally mix it with the whites of the six eggs beaten to a stiff froth.

Frances Uttinger<br>Kansas City, Kan.

## INVALID'S TEA

1 level teaspoon tea, 1 cup scalded milk. Sugar to taste

Bring milk quickly to the scalding point and pour it over the tea. Let the two infuse four minutes. Strain, and serve with or without sugar.

Tea made by this method nourishes as well as stimulates.

Frances Uttinger<br>Kansas City, Kan.

## BUTTERSCOTCH SUGAR COOKIES

Cream 2 cups brown sugar with 1/2 cup butter . Add 2 eggs, mix until creamy.

Sift together 1/2 teaspoon soda, 1/2 teaspoon cream of tartar, 3 cups flour. Mix into batter mixture, add 1 teaspoon vanilla.

Chill dough and make into little balls the size of walnuts, and flatten with fork. Bake at 350º for 8 to 10 minutes.

Frances Uttinger<br>Kansas City, Kan.

## OLD FASHIONED TOMATO SOUP

| 1 pint canned tomatoes and juice | Cut up fine whole tomatoes |
|---|---|

Bring to boil in pan. Meanwhile take another pan and heat together:

| 1 tablespoon butter | 2 tablespoon flour |
|---|---|

Stir butter and flour until smooth. Add enough milk to make thick creamy mixture. Add salt and pepper. After white sauce is made, more milk can be added for thinner soup.

To boiling tomatoes add 1/2 teaspoon soda. Pour into white sauce mixture and stir together. Serve hot.

Frances Uttinger<br>
Kansas City, Kan.

## BUTTERSCOTCH COOKIES

2 cups brown sugar
3 1/2 cups flour
1 teaspoon cream of tartar
1 cup chopped dates

2 eggs
1 cup butter
1 teaspoon baking soda
1 cup chopped pecans

Sift flour with cream of tartar and soda. Cream butter and sugar. Adding eggs, continue creaming. Add flour mixture and blend until flour disappears.

Add dates and nuts. Form in roll. Let stand overnight in cool spot, then cut in slices. Bake 8 to 10 minutes at 400°. Be careful not to burn.

"This recipe dates back to the early 1940's."

Florney Gibbons<br>
Kansas City, Mo.

## NO STIR COBBLER (Apple or Peach)

Melt one stick margarine in a baking dish. Make batter of:

1 cup flour
1 1/2 teaspoons baking powder

1 cup sugar

Mix above ingredients well. Add one egg and 3/4 cup milk.

Pour this mixture over 2 or 3 cups sliced apples or peaches. Sprinkle cinnamon over top, do not stir in. Bake at 350° about an hour.

"This is a favorite of my sister, Alene Lawson, Dekalb, Mo."

Jessie Black<br>
St. Joseph, Mo.

## SUBSTITUTIONS

I didn't have potatoes,
So I substituted rice.
I didn't have paprika,
So I added another spice.
I didn't have tomato sauce,
I used tomato paste,
A whole can, not half a can,
I don't believe in waste.
A friend gave me the recipe,
She said you couldn't beat it.
There must be something wrong here,
I couldn't even eat it!

(Submitted by)
Kathleen Potter
Savannah, Mo.

# Index of Recipes

| | |
|---|---|
| 1-2-3 Pie Pastry | 123 |
| A Healing Ointment | 45 |
| Another Cough Syrup | 45 |
| Apple Crisp | 67 |
| Apple Nut Cake | 123 |
| Apple Soup | 22 |
| Applesauce Cake | 93 |
| Aunt Ethel's Banana Nut Bread | 96 |
| Aunt Katie's Quick Indian Pudding | 110 |
| Aunt Opal's Brown Sugar Pie | 112 |
| Aunt Reba's Sugar Cookies | 103 |
| Aunty's Cole Slaw | 150 |
| Baked Apples | 144 |
| Baked Macaroni and Tomatoes | 113 |
| Banana-Carrot Nutbread | 68 |
| Beaten Biscuits | 45 |
| Beaten Biscuits | 75 |
| Beef Crust Pizza | 155 |
| Black Charley Cake | 121 |
| Black George Cake | 40 |
| Black Walnut Pie | 120 |
| Blackberry Cordial | 71 |
| Boiled Apple Dumplings | 59 |
| Boiled Peas | 24 |
| Boiled Raisin Cake | 59 |
| Bread Pudding | 96 |
| Broiled Ham & Eggs | 25 |
| Bubble Bread | 97 |
| Buckwheat Pancakes | 65 |
| Bunch of Sweet Herbs | 21 |
| Burnt Sugar Angel Food Cake | 157 |
| Butter-Less, Egg-Less, Milk-Less Cake | 103 |
| Butterscotch Cookies | 159 |
| Butterscotch Pie | 120 |
| Butterscotch Sugar Cookies | 158 |
| Cake with Nut Filling | 101 |
| Care of the Teeth | 21 |
| Carrot Salad | 147 |
| Castor Oil Cookies | 57 |
| Cement for Cracks in Stoves | 55 |
| Cheap and Easy Hot Sauce | 119 |
| Cheap Fruit Cake | 97 |
| Chicken and 'Slickers' | 67 |
| Chicken Dumplings | 122 |
| Chicken Pie | 58 |
| Chicken Pudding | 30 |
| Chiffon Cake for Blarney Stones | 118 |
| Chocolate | 23 |
| Chocolate Cake | 76 |
| Christmas Oyster Salad | 100 |
| Cider | 22 |
| Cider Cake | 27 |
| Cinnamon Waffles (Aimmet Waffles) | 69 |
| Coffee | 28 |
| Cooked Lettuce | 14 |
| Corn Chowder No. 2 | 101 |
| Corn Chowder (Soup) | 124 |
| Corn Syrup | 15 |
| Cottage Pudding | 117 |
| Country Brown Bread | 144 |
| Cranberry Jelly | 114 |
| Cranberry Pudding | 70 |
| Cranberry Salad | 151 |
| Cranberry-Walnut Relish | 78 |
| Creamed Onions | 98 |
| Custard | 60 |
| Dandelion | 28 |
| Dark Red Cake | 72 |
| Date Pudding | 69 |
| Deep Dish Pizza | 154 |
| Depression Days Cake | 145 |
| Divinity Fudge | 105 |
| Dried Apples for Pies | 28 |
| Dried Corn | 74 |
| Drop Doughnuts | 105 |
| Early-Day Cookie | 14 |
| Egg Butter | 107 |
| English Plum Pudding | 20 |
| Every Day Cookies | 109 |
| Farm Cheese | 15 |
| Fat Onions | 100 |
| Favorite Cream Cake | 114 |
| Flapper Pudding | 117 |
| Flour Pudding | 158 |
| Fresh Chopped Apple Cake | 144 |
| Fried Sweetings | 92 |
| Frozen Fruit Salad | 112 |
| German Rhubarb Pie | 116 |
| German Shortcake | 98 |
| Ginger Beer | 25 |
| Ginger Bread | 66 |
| Gingerbread | 62 |
| Graham Cracker Cookies | 78 |
| Grandma Bryan's Leppe Cookies | 66 |
| Grandma Opal's Ice Box Rolls | 61 |
| Grandma Pugh's Chow Chow | 44 |
| Grandma's Cream Puffs | 103 |
| Grandma's Delicious Sugar Cookies | 104 |
| Grandmother Osburn's Bread & Butter Pickles | 58 |
| Grandmother's Boiled Dandelions | 18 |
| Granny's Carrot Cookies | 143 |

| Recipe | Page |
|---|---|
| Great-Aunt Rosa's Watermelon Rind Preserves | 70 |
| Green Tomato Mince Pie | 153 |
| Gum Drop Cookies | 57 |
| Gumbo Soup | 22 |
| 'Hand-Me-Down' Chocolate Cake | 107 |
| Harvester Cake | 54 |
| Hermitage Bread Pudding | 92 |
| Hershey Chocolate Cake | 152 |
| Hickory Nut Macaroons | 16 |
| Holiday Cake | 126 |
| Homemade Mayonnaise | 143 |
| Homemade Noodles | 46 |
| Honey Cured Bacon | 76 |
| Hoosier Chocolate Sheet Cake | 150 |
| Hot Sweet Mustard | 56 |
| Household Remedies, Circa 1908 | 99 |
| How to Clean Mica in Stoves | 55 |
| How to Clean Vinegar Cruets | 55 |
| How to Cook a Skunk | 42 |
| Hulled Pecans | 77 |
| Invalid's Tea | 158 |
| Italian Broiled Steak Rolls | 129 |
| Italian Hot Wine Punch | 129 |
| Italian Stuffed Mushrooms | 128 |
| Italian Thumbprint Cookies | 127 |
| Italian Wine Strips | 128 |
| Jellied Veal Loaf | 112 |
| Jerusalem Pudding | 113 |
| Jessie's Refrigerator Fruit Cake | 147 |
| Judy's Chocolate Pie | 151 |
| Jumbles | 32 |
| Knee Patches | 95 |
| Korn Cake | 18 |
| Lace Cookies | 75 |
| Ladies' Cabbage | 42 |
| Lebkuchen Cookies | 62 |
| Lemon Butter | 64 |
| Lemon Pie | 157 |
| Lemon Syrup | 77 |
| Lemonade | 25 |
| Mary Ball Washington's Black Cake | 17 |
| Meat Substitute | 125 |
| Mince Meat | 116 |
| Mincemeat | 121 |
| Missouri Corn Cakes | 32 |
| Missouri Country Ham | 40 |
| Molasses Candy | 54 |
| Molasses Hand Cream (Taffy) | 77 |
| Molasses Pie | 63 |
| Mom Mellor's Lebkuchen Cookies | 67 |
| Mom's Applesauce Cake | 119 |
| Mom's Carrot Cake | 95 |
| Mush Biscuits | 109 |
| My Grandmother's Oatmeal Cookies | 97 |
| Never Fail Pie Crust | 101 |
| No Fail Divinity | 146 |
| No Stir Cobbler (Apple or Peach) | 159 |
| No Sugar French Bread | 155 |
| Norma's Brandied Peaches | 71 |
| Oatmeal Bread | 16 |
| Oatmeal Bread | 115 |
| Oatmeal Cake | 63 |
| Oatmeal Cookies | 149 |
| Old Fashioned Dumplings | 41 |
| Old Fashioned Molasses Custard Pie | 122 |
| Old Fashioned Pork Sausage | 108 |
| Old Fashioned Tomato Soup | 158 |
| Old Time Cough Cure | 44 |
| Onion Soup Au Gratin | 156 |
| Orange Slice Cake | 148 |
| Parsnip Stew | 100 |
| Peanut Butter Cookies | 115 |
| Pear Honey | 75 |
| Peas & Lettuce | 26 |
| Pepper Relish | 64 |
| Pepperoni Sausage | 145 |
| Pineapple Sheet Cake | 148 |
| Pink Arctic Freeze | 149 |
| Plum Cake | 151 |
| Plum Pudding | 60 |
| Plum Pudding | 122 |
| Poor Man's Cookies | 94 |
| Pork Sausage | 31 |
| Potato Cake | 111 |
| Potato Rolls | 43 |
| Potato Salad | 24 |
| Potato Salad | 135 |
| Prairie Pie | 46 |
| Pump 'Can' Bread | 108 |
| Pumpkin Cookies | 73 |
| Queen of Pudding | 24 |
| Quick Cake | 111 |
| Raw Apple Cake | 153 |
| Rhubarb Cake | 73 |
| Rice and Cabbage Soup | 113 |
| Roman Punch | 128 |
| Rose Brandy | 21 |
| Rose Water for Cakes & Puddings | 27 |
| Rum Catsup | 23 |
| Rum Tum Tiddy | 110 |
| Rusks | 26 |
| Sandwich Spread | 119 |
| Saucepan Oatmeal Cookies | 72 |
| Sauerkraut | 31 |
| Scrapple | 20 |

| | |
|---|---|
| Shipwreck Casserole | 59 |
| Shrove Tuesday Pancakes | 56 |
| Sicilian Spaghetti Casserole | 130 |
| Skillet Macaroni | 105 |
| Snow Ice Cream | 121 |
| Soup from Alsace | 93 |
| Sour Cream Raisin Pie | 106 |
| Sour Milk Doughnuts | 17 |
| Southern Pecan Pie | 41 |
| Special Baked Chicken | 152 |
| Spinach and Spaghetti | 149 |
| Spook Salad | 154 |
| Spoon Bread | 27 |
| Steam Pudding | 126 |
| Stew Soup | 27 |
| Stewed Carrots | 29 |
| Strawberry Biscuits | 23 |
| Stringed Beans | 30 |
| Substitutions | 160 |
| Succotash | 64 |
| Suet Pudding | 70 |
| Sweet Potato Balls | 31 |
| Sweet Sour Pork | 118 |
| Swiss Pudding | 106 |
| Switchel | 22 |
| Tea Punch | 102 |
| Tender Homemade Noodles | 146 |
| Texacus | 19 |
| Texas White Cake | 147 |
| To Make Yeast | 29 |
| To Plank a Fish | 29 |
| To Restore Gilt Frames | 55 |
| Tomato Soup | 124 |
| Tomato-Onion Salad | 136 |
| Turnip Greens with Cornmeal Dumplings | 134 |
| Vanilla Custard Whipt on the Fire | 61 |
| Vinegar Egg Pastry | 145 |
| Vinegar Icing | 56 |
| Vinegar Pie | 56 |
| Watermelon | 23 |
| Wilted Lettuce | 135 |
| Window Washing Solution | 156 |
| Yeast Biscuits | 43 |
| Yum Yum Bars (Dates) | 94 |
| Zesty Sauce for Greens | 135 |

# My Early Day Recipes

# My Early Day Recipes

800-241-1500
Piano By Candlelight
19.99 — 3 casset